The Human Brain during the First Trimester 40- to 42-mm Crown-Rump Lengths

This sixth of 15 short atlases reimagines the classic 5-volume *Atlas of Human Central Nervous System Development*. This volume presents serial sections from specimens between 40 mm and 42 mm with detailed annotations, together with 3D reconstructions. An introduction summarizes human CNS development by using high-resolution photos of methacrylate-embedded rat embryos at a similar stage of development as the human specimens in this volume. The accompanying Glossary gives definitions for all the terms used in this volume and all the others in the *Atlas*.

Features

- Classic anatomical atlas
- Detailed labeling of structures in the developing brain offers updated terminology and the identification of unique developmental features, such as, germinal matrices of specific neuronal populations and migratory streams of young neurons
- Appeals to neuroanatomists, developmental biologists, and clinical practitioners
- A valuable reference work on brain development that will be relevant for decades

ATLAS OF
HUMAN CENTRAL NERVOUS SYSTEM DEVELOPMENT
Series

Volume 1: The Human Brain during the First Trimester 3.5- to 4.5-mm Crown-Rump Lengths

Volume 2: The Human Brain during the First Trimester 6.3- to 10.5-mm Crown-Rump Lengths

Volume 3: The Human Brain during the First Trimester 15- to 18-mm Crown-Rump Lengths

Volume 4: The Human Brain during the First Trimester 21- to 23-mm Crown-Rump Lengths

Volume 5: The Human Brain during the First Trimester 31- to 33-mm Crown-Rump Lengths

Volume 6: The Human Brain during the First Trimester 40- to 42-mm Crown-Rump Lengths

Volume 7: The Human Brain during the First Trimester 57- to 60-mm Crown-Rump Lengths

Volume 8: The Human Brain during the Second Trimester 96- to 150-mm Crown-Rump Lengths

Volume 9: The Human Brain during the Second Trimester 160- to 170-mm Crown-Rump Lengths

Volume 10: The Human Brain during the Second Trimester 190- to 210-mm Crown-Rump Lengths

Volume 11: The Human Brain during the Third Trimester 225- to 235-mm Crown-Rump Lengths

Volume 12: The Human Brain during the Third Trimester 260- to 270-mm Crown-Rump Lengths

Volume 13: The Human Brain during the Third Trimester 310- to 350-mm Crown-Rump Lengths

Volume 14: The Spinal Cord during the First Trimester

Volume 15: The Spinal Cord during the Second and Third Trimesters and the Early Postnatal Period

The Human Brain during the First Trimester 40- to 42-mm Crown-Rump Lengths

Atlas of Human Central Nervous System Development, Volume 6

Shirley A. Bayer
Joseph Altman

CRC Press
Taylor & Francis Group
Boca Raton London New York

CRC Press is an imprint of the
Taylor & Francis Group, an **informa** business

First edition published 2023
by CRC Press
6000 Broken Sound Parkway NW, Suite 300, Boca Raton, FL 33487-2742

and by CRC Press
4 Park Square, Milton Park, Abingdon, Oxon, OX14 4RN

CRC Press is an imprint of Taylor & Francis Group, LLC

LCCN 2022008216

ISBN: 978-1-032-18334-3 (hbk)
ISBN: 978-1-032-21937-0 (pbk)
ISBN: 978-1-003-27065-2 (ebk)

DOI: 10.1201/9781003270652

Publisher's note: This book has been prepared from camera-ready copy provided by the authors.

Access the Support Material: www.routledge.com/9781032183343

Typeset in Times Roman by KnowledgeWorks Global Ltd.

CONTENTS

ACKNOWLEDGMENTS

We thank the late Dr. William DeMyer, pediatric neurologist at Indiana University Medical Center, for access to his personal library on human CNS development. We also thank the staff of the National Museum of Health and Medicine that were at the Armed Forces Institute of Pathology, Walter Reed Hospital, Washington, D.C. when we collected data in 1995 and 1996: Dr. Adrianne Noe, Director; Archibald J. Fobbs, Curator of the Yakovlev Collection; Elizabeth C. Lockett; and William Discher. We are most grateful to the late Dr. James M. Petras at the Walter Reed Institute of Research who made his darkroom facilities available so that we could develop all the photomicrographs on location rather than in our laboratory in Indiana. Finally, we thank Chuck Crumly, Neha Bhatt, and Kara Roberts, Michele Dimont, and Rebecca Condit for expert help during production of the manuscript.

AUTHORS

Shirley A. Bayer received her PhD from Purdue University in 1974 and spent most of her scientific career working with Joseph Altman. She was a professor of biology at Indiana-Purdue University in Indianapolis for several years, where she taught courses in human anatomy and developmental neurobiology while continuing to do research in brain development. Her lengthy publication record of dozens of peer-reviewed, scientific journal articles extends back to the mid 1970s. She has co-authored several books and many articles with her late spouse, Joseph Altman. It was her research (published in *Science* in 1982) that proved that new neurons are added to granule cells in the dentate gyrus during adult life, a unique neuronal population that grows. That paper stimulated interest in the dormant field of adult neurogenesis.

Joseph Altman, now deceased, was born in Hungary and migrated with his family via Germany and Australia to the US. In New York, he became a graduate student in psychology in the laboratory of Hans-Lukas Teuber, earning a PhD in 1959 from New York University. He was a postdoctoral fellow at Columbia University, and later joined the faculty at the Massachusetts Institute of Technology. In 1968, he accepted a position as a professor of biology at Purdue University. During his career, he collaborated closely with Shirley A. Bayer. From the early 1960s-2016, he published many articles in peer-reviewed journals, books, monographs, and free online books that emphasized developmental processes in brain anatomy and function. His most important discovery was adult neurogenesis, the creation of new neurons in the adult brain. This discovery was made in the early 1960s while he was based at MIT, but was largely ignored in favor of the prevailing dogma that neurogenesis is limited to prenatal development. After Dr. Bayer's paper proved new neurons are added to granule cells in the hippocampus, Dr. Altman's monumental discovery became more accepted. During the 1990s, new researchers "rediscovered" and confirmed his original finding. Adult neurogenesis has recently been proven to occur in the dentate gyrus, olfactory bulb, and striatum through the measurement of Carbon-14—the levels of which changed during nuclear bomb testing throughout the 20th century—in postmortem human brains. Today, many laboratories around the world are continuing to study the importance of adult neurogenesis in brain function. In 2011, Dr. Altman was awarded the Prince of Asturias Award, an annual prize given in Spain by the Prince of Asturias Foundation to individuals, entities, or organizations globally who make notable achievements in the sciences, humanities, and public affairs. In 2012, he received the International Prize for Biology - an annual award from the Japan Society for the Promotion of Science (JSPS) for "outstanding contribution to the advancement of research in fundamental biology." This Prize is one of the most prestigious honors a scientist can receive. When Dr. Altman died in 2016, Dr. Bayer continued the work they started over 50 years ago. In her late husband's honor, she created the Altman Prize, awarded each year by JSPS to an outstanding young researcher in developmental neuroscience.

INTRODUCTION

ORGANIZATION OF THE ATLAS

This is the sixth book in the *Atlas of Human Central Nervous System Development* series, 2nd Edition. It deals with human brain development in three normal specimens during the middle first trimester with crown-rump (CR) lengths from 40- to 42-mm and estimated gestation weeks (GW) from 9.5 to 9.6 (Loughna et al., 2009). These specimens were analyzed in Volume 4 of the 1st Edition (Bayer and Altman, 2006). One specimen (M841) is from the *Minot Collection.*[1] The other two (C886 and C6658) are from the *Carnegie Collection.*[2] Both collections are now in the National Museum of Health and Medicine, which used to be located at the Armed Forces Institute of Pathology (AFIP) in Walter Reed Hospital in Washington, D.C. When the AFIP closed, the National Museum moved to Silver Springs, MD; this collection is still available for research. M841 is cut in the frontal/horizontal plane, C886

in the horizontal plane, and C6658 is cut in the sagittal plane. The three section planes in specimens of the same age give a more complete perspective of the structure of the brain at this time. As in the previous volumes of the *Atlas*, each specimen is presented in serial grayscale photographs of its Nissl-stained sections showing the brain and surrounding tissues (**Parts II–IV**). The photographs are shown from anterior to posterior (frontal/horizontal specimen), dorsal to ventral (horizontal specimen), and medial to lateral (sagittal specimen). The dorsal part of each frontal/horizontal section is toward the top of the page, the ventral part at the bottom, and the midline is in the vertical center. In the horizontal specimen, the left side of the section is anterior, right side, posterior, and the midline is in the horizontal center. In the sagittal specimen, the left side of each section is anterior, right side posterior, top side dorsal, and bottom side ventral.

PLATE PREPARATION

All sections of a given specimen were photographed at the same magnification. Sections throughout the entire specimen were photographed in serial order with Kodak technical pan black-and-white negative film (#TP442). The film was developed for 6 to 7 minutes in dilution F of Kodak HC-110 developer, stop bath for 30 seconds, Kodak fixer for 5 minutes, Kodak hypo-clearing agent for 1 minute, running water rinse for 10 minutes, and a brief rinse in Kodak photo-flo before drying. Negatives were scanned as color positives at 2700 dots per inch (dpi) with a Nikon Coolscan-1000 35-mm negative film scanner attached to a Macintosh PowerMac G3 computer which had a plug-in driver built into Adobe Photoshop. Images were converted to 300 dpi using the non-resampling method for image size. Using the powerful features of Adobe photo-

---

1. The *Minot Collection* (designated by an **M** prefix in the specimen number) is the work of Dr. Charles Sedgwick Minot (1852–1914), an embryologist at Harvard University. Throughout his career, Minot collected about 1900 embryos from a variety of species. The 100 human embryos in the group were probably acquired between 1900 and 1910. From our examination of these specimens and their similar appearance, we assume that they are preserved in the same way, although we could not find any records describing fixation procedures. The slides contain information on section numbers, section thickness (6 μm or 10 μm), and stain (aluminum cochineal).

2. The *Carnegie Collection* (designated by a **C** prefix in the specimen number) started in the Department of Embryology of the Carnegie Institution of Washington. It was led by Franklin P. Mall (1862–1917), George L. Streeter (1873–1948), and George W. Corner (1889–1981). These specimens were collected during a span of 40 to 50 years and were histologically prepared with a variety of fixatives, embedding media, cutting planes, and histological stains. Early analyses of specimens were published in the early 1900s in *Contributions to Embryology, The Carnegie Institute of Washington* (now archived in the Smithsonian Libraries). O'Rahilly and Müller (1987, 1994) have given overviews of some first trimester specimens in this collection.

shop, contrast was enhanced, uneven staining was corrected, and areas of uneven exposure were slightly darkened or lightened.

The photos chosen for annotation in **Parts II–IV** are presented as companion plates. The *low-magnification plates* of the frontal/horizontal and horizontal specimens are designated as **A** and **B** on one set of facing pages. **Part A** on the left shows the full-contrast photo, while **Part B** on the right shows a low-contrast copy with annotations. Plates of the sagittal specimen are designated as **A** through **D** on two sets of facing pages. **Part A** on the left page shows the full-contrast photograph of the brain in the skull without labels. **Part B** on the right page shows low-contrast copies of the same photograph with superimposed outlines of brain parts and labels of major brain ventricles and structures. **Part C** on the second left page shows a full-contrast photo of a slightly larger brain "dissected" from its peripheral structures, except cranial sensory structures have been preserved. **Part D** on the second right page is a low-contrast copy of the photo in C with more detailed labeling in the brain. Several *high-magnification plates* feature enlarged views of the brain to show tissue organization. This allows users to see the entire section and then consult the detailed markup in the low-contrast copy on the facing page, leaving little doubt about what is being identified. The labels themselves are not abbreviated, so there is no lookup on a list. Different fonts are used to label different classes of structures: the ventricular system is labeled in **CAPITALS**, the neuroepithelium and other germinal zones in **Helvetica bold**, transient structures in ***Times bold italic***, and permanent structures in Times Roman or **Times bold**. Using Adobe Illustrator, labels were superimposed and lines were drawn around structural details on the low-contrast images. Plates were placed into a book layout using Adobe InDesign. Finally, high-resolution portable document files (pdf) were uploaded to CRC Press/Taylor & Francis websites.

DEVELOPMENT IN SPECIMENS
(CR 40–42-mm)

The specimens in this volume are equivalent to rat embryos on embryonic day (E) 18 based on our morphological matching. E18 rats have a similar appearance to human specimens from 40- to 42-mm crown-rump lengths. Our timetables of neurogenesis use ^{3}H-thymidine dating methods (Bayer and Altman, 1991, 1995, 2012-present; Bayer et al., 1993, 1995) to determine neuronal populations that are being generated in E18 rats; we assume that is comparable to neurogenesis in 40- to 42-mm human specimens (Bayer et al., 1993, 1995; Bayer and Altman, 1995). **Table 1** lists populations being generated throughout the neuraxis: the brainstem and midbrain tectum (**Table 1A**), the diencephalon (**Table 1B**), the pallidum/striatum, amygdala, and septum (**Table 1C**), and the cerebral cortex, hippocampus, and olfactory structures (**Table 1D**). We use photos of methacrylate-embedded rat embryos on E18 (Bayer,

2013-present) to show the fine details of immature structures in the medulla, pons, cerebellum, superior colliculus, cerebral cortex and olfactory bulb (**Figs. 1–8**) because the preservation of human specimens does not show detail.

The structures we are showing in the rat embryos are some of the immature features in the brain at this time. In the medulla, young neurons in the posterior extramural migratory stream are still crossing the midline (**Fig. 1**), although in fewer numbers than shown in the specimens analyzed in Volume 5 (Bayer and Altman, in press). Many of these neurons (arising in the precerebellar NEP in the dorsal medulla) have already settled in the lateral reticular and external cuneate nuclei. But the precerebellar NEP is still the most active germinal zone in the medulla (**Fig. 2**) and is now generating neurons that will migrate to the pons and settle in the pontine nuclei. The pons does not show settling neurons in the pontine nuclei, but they are on their way in the anterior extramural migratory stream, a definite structure on the anterior surface (**Fig. 3**).

The cerebellum (**Fig. 4**) is greatly expanding. Many deep nuclear neurons are leaving superficial parts to settle beneath the growing cortex. The cortical area is always limited to the part just beneath the mitotically active external germinal layer (egl), which is one prong of a germinal trigone. Purkinje cells are accumulating beneath the egl and some fibers are already in a primordial molecular layer. The cerebellar NEP itself is still mitotically active and is making progenitors of Golgi cells that will eventually disperse throughout the granule cell layer.

Another immature feature is in the tectum of the midbrain. We illustrate the evolving layering in the superior colliculus (**Fig. 5**), where lens-like bundles of axons are surrounded by migrating neurons in a honeycomb matrix. It is our hypothesis that young neurons intermingle and have a "handshake" with the axons that will innervate these neurons in the mature brain.

The evolving layering of the neocortex (**Fig. 6**) in the telencephalon shows a thick cortical plate throughout its entire extent. Cajal-Retzius neurons are prominent in layer I, while subplate neurons are delaminating from the lower parts of the cortical plate and are intermingling with fibers in the glial channels at the base of the plate. A wide swath of migrating neurons lies beneath the subplate—the stratified transitional field (STF), where pyramidal cells that will reside in various layers of the mature neocortex sojourn to intermingle with incoming axons from the thalamus and other parts of the brain. Indeed, the STF is an important staging area for the maturation of cortical circuitry.

The hippocampus is another immature telencephalic structure (**Fig. 7**) where a few neurons are settling in the CA fields of Ammon's horn. Most likely, these will reside in field CA3. The thick hippocampal NEP has a notched

Table 1A: Neurogenesis by Region

REGION and NEURAL POPULATION	CROWN RUMP LENGTH 40-42 mm
PRECEREBELLAR NUCLEI	
Pontine nuclei	● ●
SUPERIOR COLLICULUS	
stratum album	●
stratum lemnisci	●
stratum griseum intermediate	●
stratum griseum superficial	●
INFERIOR COLLICULUS	
Anterior intermediate	● ●
Posterior intermediate	● ●
Anteromedial	● ●
Posteromedial	●

Table 1A. Neural populations in the precerebellar nuclei, mesencephalic tegmentum, superior colliculus, and inferior colliculus that are being generated in rats on Embryonic day (E) 17 (comparable to humans at CR 31-33-mm). *Green dots* indicate the amount of neurogenesis occurring: one dot=<15%; two dots=15-90%. This same dot notation is used for all of the remaining parts (**B-E**) of **Table 1**.

Table 1B: Neurogenesis by Region

REGION and NEURAL POPULATION	CROWN RUMP LENGTH 40-42 mm
PREOPTIC AREA/ HYPOTHALAMUS	
Medial preoptic nucleus	●
Sexually dimorphic nucleus	● ●
Periventricular preoptic nucleus	● ●
Median preoptic nucleus	●
Arcuate nucleus	● ●
Tuberommillary nucleus	●
Medial mammillary n. (ventral)	●
THALAMUS/EPITHALAMUS	
Paraventricular	●
Paratenial	●
Medial habenula	● ●

Table 1C: Neurogenesis by Region

REGION and NEURAL POPULATION	CROWN RUMP LENGTH 40-42 mm
PALLIDUM AND STRIATUM	
Olfactory tubercle (small neurons)	●
Caudate and putamen	●
Nucleus accumbens	●
Islands of Calleja	●
AMYGDALA	
Central nucleus	●
Intercalated masses	●
Amygdalo-hippocampal area	● ●
SEPTUM	
Lateral nucleus	●

Table 1D: Neurogenesis by Region

REGION and NEURAL POPULATION	CROWN RUMP LENGTH 40-42 mm
NEOCORTEX and LIMBIC CORTEX	
Layer V	●
Layers IV-II	● ●
OLFACTORY CORTEX	
Layer II (anterior)	●
Layer II (posterior)	●
Layers III-IV (anterior)	●
HIPPOCAMPAL REGION	
Entorhinal cortex Layer III	●
Subiculum (deep)	● ●
Subiculum (superficial)	● ●
Ammon's Horn CA1	● ●
Ammon's Horn CA3	● ●
Ammon's Horn CA4	● ●
OLFACTORY BULB	
Internal tufted cells (main bulb)	●
External tufted cells (main bulb)	● ●
ANTERIOR OLFACTORY NUCLEUS	
Pars externa	●
AON proper	● ●

curve that extends downward to connect with the choroid plexus. Part of that germinal zone is probably a glioepithelium that will generate oligodendroglia in the fimbria, but most of it is the source of the dentate migration. Many cells are migrating from the NEP in the "dentate notch" and accumulate beneath Ammon's horn. This NEP is the source of a dispersed germinal matrix that will produce granule cells in the dentate gyrus. A clump of cells is already visible on E18. The dentate migration is visible in a rudimentary form in the specimens analyzed in Volume 4 of this series (Bayer and Altman, in press).

The main developmental event in the olfactory structures is the mitral cell migration into the olfactory bulb (**Fig. 8**).

These cells appear to enter the bulb from the lateral side, where their trailing axons accumulate in the lateral olfactory tract. Once inside, migrating cells fan out and peel off in curved arrays to settle in the primordial mitral cell layer throughout the entire extent of the main olfactory bulb.

All terms used in the tables, figures, and plates identify maturing and unique developmental structures that are often unfamiliar or have long been forgotten from courses in medical and graduate school. To refresh memories, please consult the extensive definitions of these terms in the *Glossary* (Bayer, in press) that accompanies the *Atlas*.

THE LOWER MEDULLA IN AN E18 RAT EMBRYO

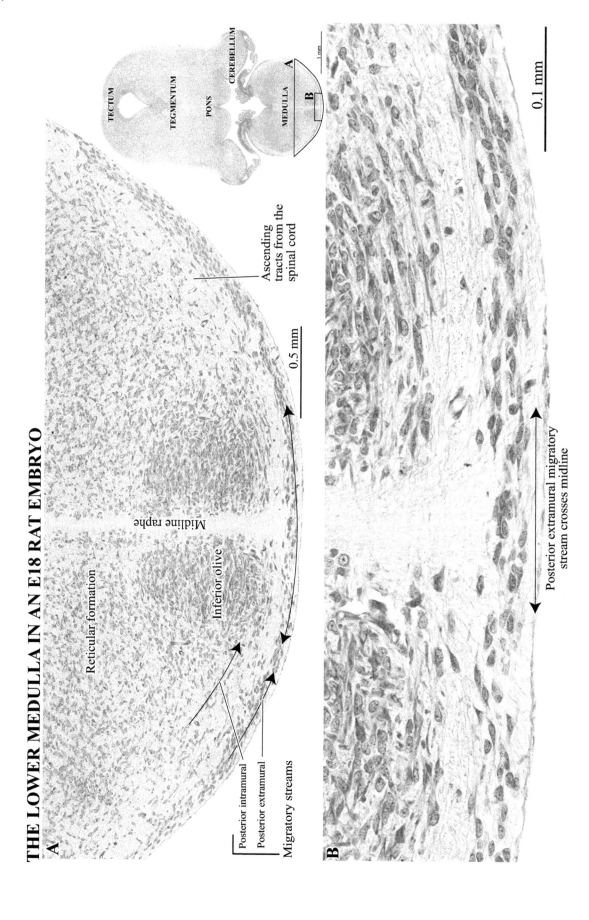

Figure 1. Frontal/horizontal section of the posterior brain in an E18 rat embryo at a similar stage as human specimens in this volume. The two posterior migratory streams (*arrows*) of precerebellar nuclear neurons are visible. The waning intramural stream takes a trajectory through the medullary parenchyma and heads for the inferior olive. The extramural stream runs beneath the pia meninx on the outer edge of the medulla. Neurons cross the midline beneath the inferior olive and will settle on the opposite side in the lateral reticular and external cuneate precerebellar nuclei. (3μ methacrylate section, toluidine blue stain). Source: braindevelopmentmaps.org (E16 coronal archive)

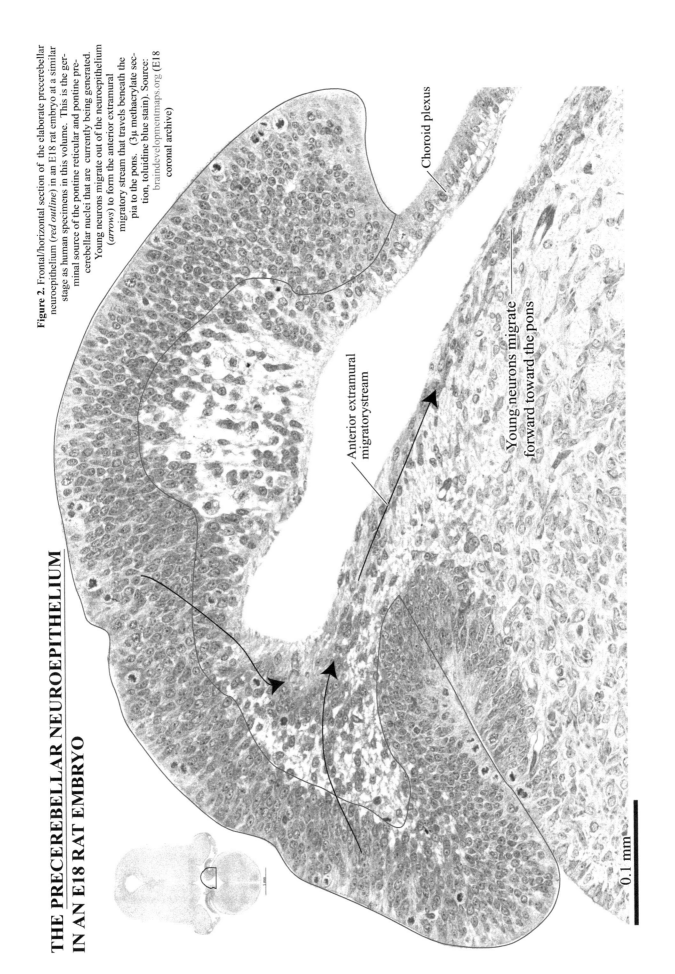

THE PRECEREBELLAR NEUROEPITHELIUM IN AN E18 RAT EMBRYO

Figure 2. Frontal/horizontal section of the elaborate precerebellar neuroepithelium (*red outline*) in an E18 rat embryo at a similar stage as human specimens in this volume. This is the germinal source of the pontine reticular and pontine precerebellar nuclei that are currently being generated. Young neurons migrate out of the neuroepithelium (*arrows*) to form the anterior extramural migratory stream that travels beneath the pia to the pons. (3μ methacrylate section, toluidine blue stain). Source: braindevelopmentmaps.org (E18 coronal archive)

Choroid plexus

Anterior extramural migratory stream

Young neurons migrate forward toward the pons

0.1 mm

THE ANTERIOR PONTINE SURFACE
IN AN E18 RAT EMBRYO

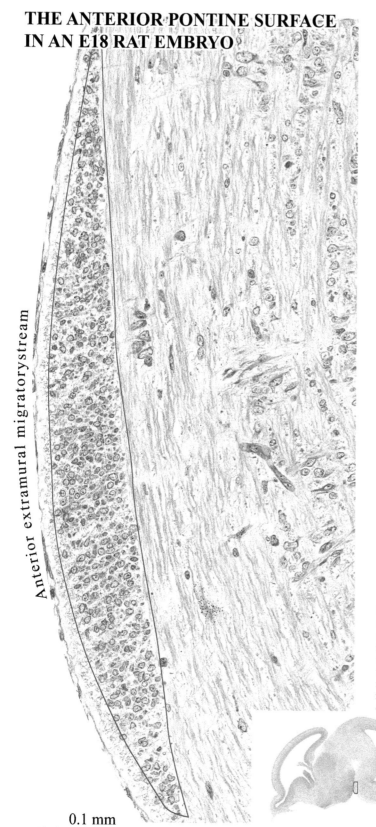

Anterior extramural migratory stream

0.1 mm

Figure 3. The anterior extramural migratory stream (*red outline*) is a lens-shaped structure on the anterior pontine surface at the base of the mesencephalic flexure. The neurons are migrating perpendicular to the section plane, so they are rounded rather than spindle-shaped. These cells will settle in the core of the large pontine precerebellar nuclei at the base of the pons in the mature brain. (3μ methacrylate section, toluidine blue stain). Source: brain-developmentmaps.org (E17 coronal archive)

THE CEREBELLUM
IN AN E18 RAT EMBRYO

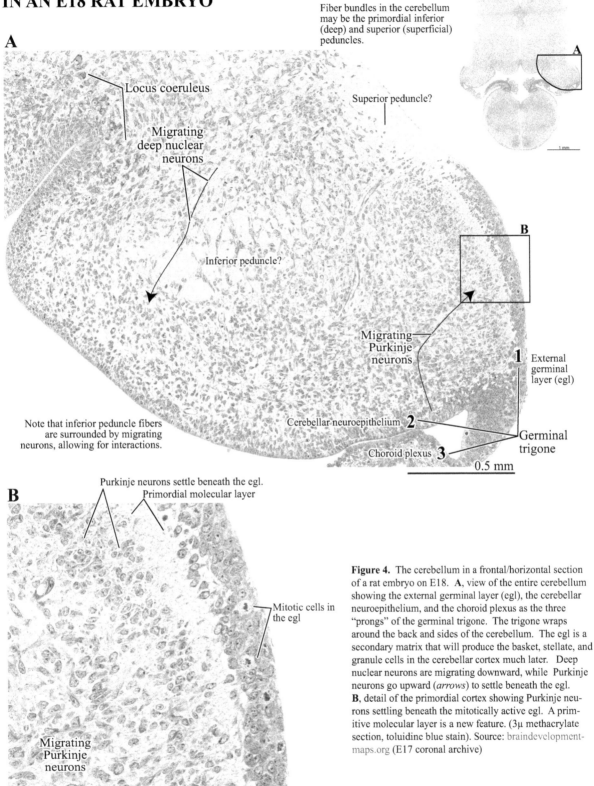

Fiber bundles in the cerebellum
may be the primordial inferior
(deep) and superior (superficial)
peduncles.

A

Locus coeruleus

Migrating
deep nuclear
neurons

Superior peduncle?

B

Inferior peduncle?

Migrating
Purkinje
neurons

1 External
germinal
layer (egl)

Cerebellar neuroepithelium **2**

Germinal
trigone

Note that inferior peduncle fibers
are surrounded by migrating
neurons, allowing for interactions.

Choroid plexus **3**

0.5 mm

B

Purkinje neurons settle beneath the egl.
Primordial molecular layer

Mitotic cells in
the egl

Migrating
Purkinje
neurons

0.1 mm

Figure 4. The cerebellum in a frontal/horizontal section
of a rat embryo on E18. **A**, view of the entire cerebellum
showing the external germinal layer (egl), the cerebellar
neuroepithelium, and the choroid plexus as the three
"prongs" of the germinal trigone. The trigone wraps
around the back and sides of the cerebellum. The egl is a
secondary matrix that will produce the basket, stellate, and
granule cells in the cerebellar cortex much later. Deep
nuclear neurons are migrating downward, while Purkinje
neurons go upward (*arrows*) to settle beneath the egl.
B, detail of the primordial cortex showing Purkinje neurons settling beneath the mitotically active egl. A prim-
itive molecular layer is a new feature. (3μ methacrylate
section, toluidine blue stain). Source: braindevelopment-
maps.org (E17 coronal archive)

8

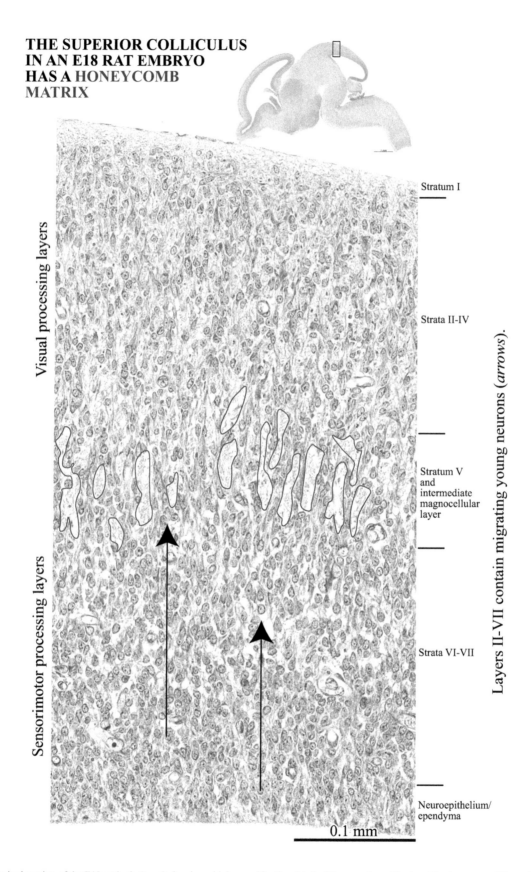

**THE SUPERIOR COLLICULUS
IN AN E18 RAT EMBRYO
HAS A HONEYCOMB
MATRIX**

Visual processing layers

Sensorimotor processing layers

Stratum I

Strata II-IV

Stratum V
and
intermediate
magnocellular
layer

Strata VI-VII

Neuroepithelium/
ependyma

Layers II-VII contain migrating young neurons (*arrows*).

0.1 mm

Figure 5. Sagittal section of the E18 rat brain (*inset*) showing a high magnification detail of the superior colliculus. The layers are still poorly defined, but superficial (I-IV) and deep layers (VI-VII) are tentatively identified by the honeycomb matrix (*red outlines* of fiber bundles) in layer V and the intermediate magnocellular layer. We postulate that the honeycomb matrix is a region where incoming axons interact with young superior colliculus neurons, a staging area for connections in the mature brain. (3μ methacrylate section, toluidine blue stain) Source: braindevelopmentmaps.org (E18 sagittal archive)

THE NEOCORTEX
IN AN E18 RAT EMBRYO

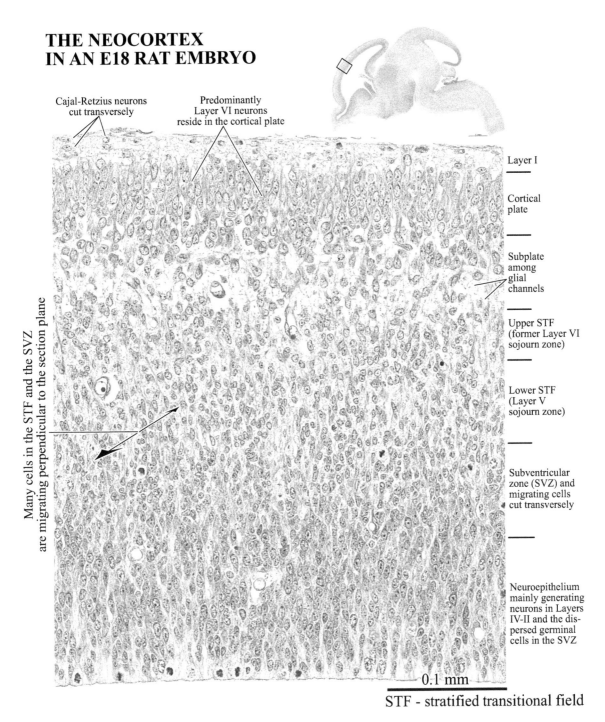

Cajal-Retzius neurons
cut transversely

Predominantly
Layer VI neurons
reside in the cortical plate

Many cells in the STF and the SVZ
are migrating perpendicular to the section plane

Layer I

Cortical
plate

Subplate
among
glial
channels

Upper STF
(former Layer VI
sojourn zone)

Lower STF
(Layer V
sojourn zone)

Subventricular
zone (SVZ) and
migrating cells
cut transversely

Neuroepithelium
mainly generating
neurons in Layers
IV-II and the dis-
persed germinal
cells in the SVZ

0.1 mm

STF - stratified transitional field

Figure 6. Sagittal section of the E18 rat brain (*inset*) showing high-magnification detail of the neocortex. In the frontal/horizontal section plane, many neurons are migrating parallel and appear as long, spindle-shaped cells. Here, all cells appear to be migrating radially, but that is because the spindle-shaped cells are being cut transversely (*white-outlined arrow*). This cortex is in the middle of the ventrolateral (more mature) to dorsomedial (less mature) region. The predominant occupants of the cortical plate are layer VI neurons; subplate neurons have mostly delaminated from the cortical plate and are mingling with the glial channels beneath it. Layer V neurons sojourn deep in the STF, but many are migrating upward toward the cortical plate, where they will stack up above the layer VI neurons. Layers IV-II neurons are being generated in the primary cortical neuroepithelium and in the subventricular zone. Eventually, they will sojourn in the deepest part of the STF. The STF is the area where the initial layout of cortical circuitry is being established with other parts of the brain. (3μ methacrylate section, toluidine blue stain)

Source: braindevelopmentmaps.org (E18 sagittal archive)

THE HIPPOCAMPUS IN AN E18 RAT EMBRYO

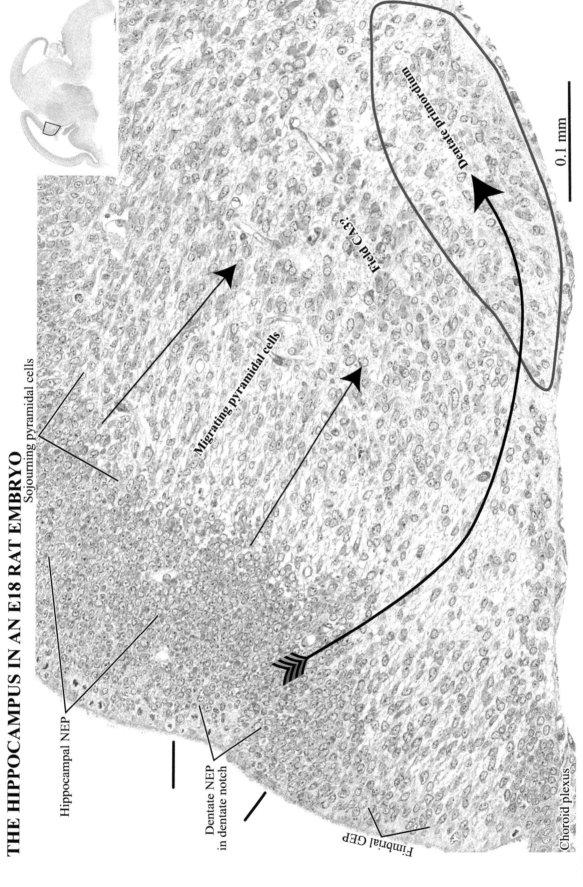

Figure 7. The hippocampus in a sagittal section of an E18 rat embryo (*inset*). The thick hippocampal neuroepithelium (NEP) is adjacent to the dentate NEP; the fimbrial glioepithelium (GEP) thins and spreads out among fibers that may be afferents from the septum. The hippocampal NEP has a pyramidal cell sojourn zone basally, and many cells are migrating from there (*straight arrows*) into primordial field CA3. There is a stream of cells (the dentate migration, *curved arrow*) heading into the dentate primordium (*red outline*). (3 μ methacrylate section, toluidine blue stain) Source: braindevelopmentmaps.org (E18 sagittal archive)

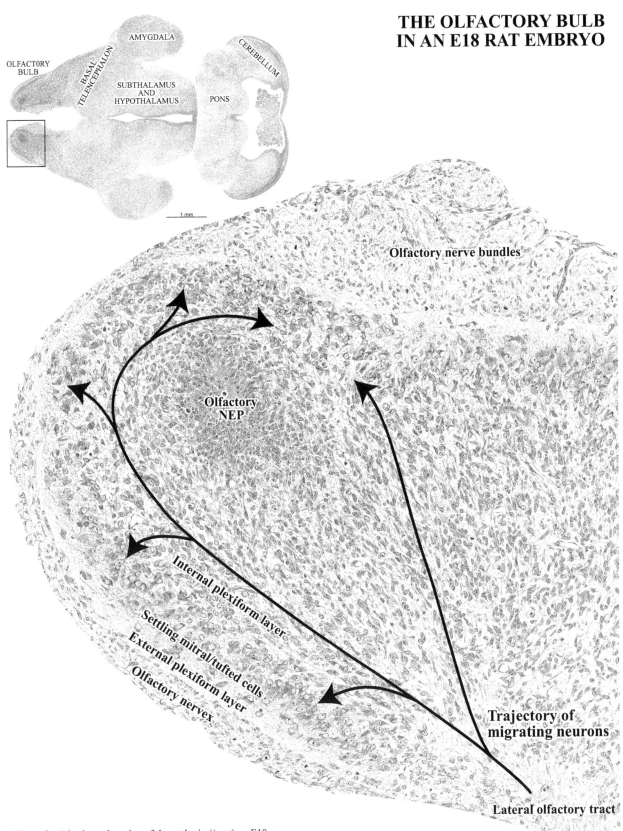

Figure 8. A horizontal section of the rat brain (*inset*) on E18 showing detail of the olfactory bulb. The mitral and tufted cell migrations come in from the lateral basal telencephalon and enter the internal plexiform layer to peel off and settle in the primordial mitral/tufted cell layer (*curved arrow with branches*). (3μ methacrylate section, toluidine blue stain) Source: braindevelopmentmaps.org (E17 coronal archive)

REFERENCES

Bayer SA, Altman J (1997) *Development of the Cerebellar System in Relation to Its Evolution, Structure, and Function*, CRC Press, Boca Raton, FL.

Bayer, SA (2013-present) www.braindevelopmentmaps. org Laboratory of Developmental Neurobiology, Ocala FL. (This website is an image database of methacrylate-embedded normal rat embryos and paraffin-embedded rat embryos exposed to ³H-Thymidine.)

Bayer, SA (in press) *Glossary to Accompany Atlas of Human Central Nervous System Development.* Taylor & Francis/CRC Press.

Bayer SA, Altman J (1991) *Neocortical Development*, Raven Press, New York.

Bayer SA, Altman J (1995) Development: Some principles of neurogenesis, neuronal migration and neural circuit formation. In: *The Rat Nervous System*, 2nd Edition, George Paxinos, Ed. Academic Press, Orlando, Florida., pp. 1079-1098.

Bayer SA, Altman J (2012-present) www.neurondevelopment.org (This website has downloadable pdf files of our scientific papers on rat brain development grouped by subject.)

Bayer SA, Altman J (2006) *Atlas of Human Central Nervous System Development*, Volume 4: *The Human Brain during the Late First Trimester.* CRC Press.

Bayer SA, Altman J (in press) *The Human Brain during the First Trimester 21- to 23-mm Crown Rump Lengths. Atlas of Human Central Nervous System Development*, Volume 5, Taylor & Francis/CRC Press.

Bayer SA, Altman J (in press) *Glossary to Accompany Atlas of Human Central Nervous System Development*, Taylor & Francis/CRC Press.

Bayer SA, Altman J, Russo RJ, Zhang X (1993) Timetables of neurogenesis in the human brain based on experimentally determined patterns in the rat. *Neurotoxicology* **14**: 83-144.

Bayer SA, Altman J, Russo RJ, Zhang X (1995) Embryology. In: *Pediatric Neuropathology*, Serge Duckett, Ed. Williams and Wilkins, pp. 54-107.

Hochstetter F (1919) *Beiträge zur Entwicklungsgeschichte des menschlichen Gehirns.* Vol. 1. Leipzig und Wien: Deuticke.

Loughna P, Citty L, Evans T, Chudleigh T (2009) Fetal size and dating: Charts recommended for clinical obstetric practice, *Ultrasound*, 17:161-167.

O'Rahilly R, Müller F. (1987) *Developmental Stages in Human Embryos, Carnegie Institution of Washington*, Publication 637.

O'Rahilly R; Müller F. (1994) *The Embryonic Human Brain*, Wiley-Liss, New York.

PART II: M841
CR 42 mm (GW 10.6)
Frontal/Horizontal

This specimen is number 841 in the Minot Collection, referred to here as M841, with a crown rump length (CR) of 42 mm, and estimated to be at gestational week (GW) 10.6. The fetus was embedded in paraffin, cut in 10 μm thick sections, and stained with borax carmine and Lyon's blue. No information is available on date of collection (sometime between 1900 and 1910) and the fixative used. Since there is no photograph of this brain before it was embedded and cut, a specimen from Hochstetter (1919) that is comparable to M841 has been modified to show the approximate section plane and external features of the brain at GW10.6 (**Fig. 9**). Like most of the specimens in this volume, the sections are not cut exactly in one plane; M841 is midway between frontal and horizontal. Photographs of 22 sections (**Plates 1-22**) are illustrated at low magnification, showing excellent tissue detail. Unfortunately, a fine granular precipitate is visible at high magnification.

In the cerebral cortex, the neuroepithelium is prominent and there is probably a small subventricular zone present (not visible because of the inability to see detail at high magnification). There is a stratified transitional field (STF), with layers we designate as STF1, STF5, and STF4. The ventroateral-to-dorsomedial maturation gradient in the cortical plate and STF is evident. Streams of neurons and glia appear to leave STF4 and enter the lateral migratory stream. A massive neuroepithelium/subventricular zone overlies the amygdala, nucleus accumbens, and striatum (caudate and putamen) where neurons (and glia) are being generated. The olfactory evagination is at the junction of the basal telencephalon and the cerebral cortex.

The diencephalic neuroepithelia have thinned out because nearly all neuronal populations are either near their last days of generation or have already been generated. Nuclear borders are indistinct throughout the diencephalon because most neurons are migrating, blurring the boundaries.

The midbrain tegmentum, pons, and medulla are lined by a thin neuroepithelium being transformed to an ependyma because the majority of neurons have been produced. Neurons throughout this part of the brain are still migrating, but many have settled, making some nuclear groups more definite. There is a prominent germinal zone in the mesencephalic tectum, where many neuronal populations are having their final days of generation. Some layering in the tectum is beginning to differentiate.

The thick precerebellar neuroepithelium is an exception in the medulla because it is still generating neurons. The anterior extramural and posterior extramural migratory streams are subpial accumulations in the medulla and pons.

The cerebellum has a definite, but thinning neuroepithelium at the ventricular surface. Probably most of the Purkinje neurons are either sojourning in a thick dense layer outside the neuroepithelium, while others are migrating upward to form a primordial layer beneath the external germinal layer (egl) that partially covers the cerebellum. The cortex is just beginning to form beneath the parts covered by the egl.

GW10.6 FRONTAL/HORIZONTAL SECTION PLANES

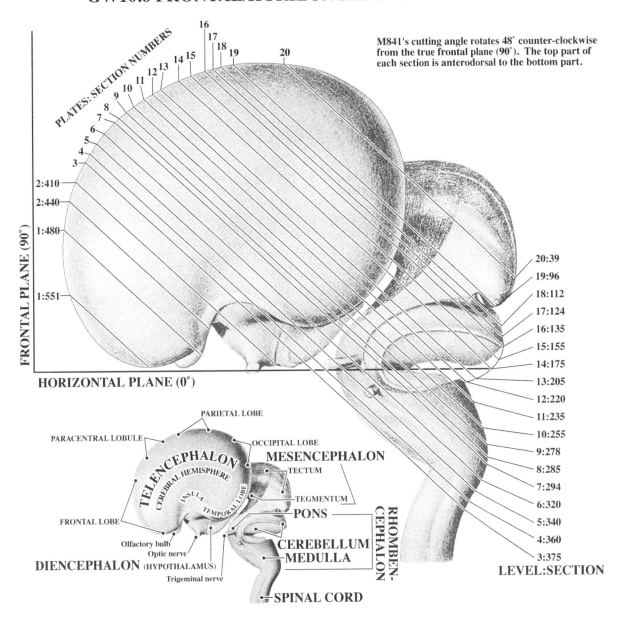

M841's cutting angle rotates 48° counter-clockwise from the true frontal plane (90°). The top part of each section is anterodorsal to the bottom part.

PLATES: SECTION NUMBERS

16 17 18 19 20
14 15
9 10 11 12 13
8
7
6
5
4
3

FRONTAL PLANE (90°)

2:410
2:440
1:480
1:551

HORIZONTAL PLANE (0°)

20:39
19:96
18:112
17:124
16:135
15:155
14:175
13:205
12:220
11:235
10:255
9:278
8:285
7:294
6:320
5:340
4:360
3:375

LEVEL:SECTION

PARIETAL LOBE
PARACENTRAL LOBULE
OCCIPITAL LOBE
TELENCEPHALON
CEREBRAL HEMISPHERE
MESENCEPHALON
TECTUM
INSULA
TEMPORAL LOBE
TEGMENTUM
FRONTAL LOBE
PONS
Olfactory bulb
Optic nerve
CEREBELLUM
MEDULLA
RHOMBEN-CEPHALON
DIENCEPHALON (HYPOTHALAMUS)
Trigeminal nerve
SPINAL CORD

Figure 9. The lateral view of the brain and upper cervical spinal cord from a specimen with a crown rump length of 38 mm (modified from Figure 43, Table VII, Hochstetter, 1919) serves to show the approximate locations and cutting angles of the illustrated sections of M841 in the following pages. The small inset identifies the major structural features. The line in the cerebellum and dorsal edges of the pons and medulla is the cut edge of the medullary velum.

PLATE 1A
CR 42 mm,
GW 10.6, M841
Frontal/Horizontal

Section 551

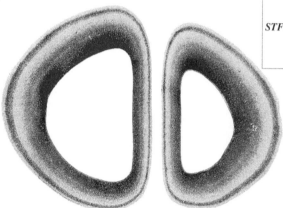

LAYERS OF THE CORTICAL
STRATIFIED TRANSITIONAL FIELD (STF)

STF1 Superficial fibrous layer with an early developmental stage *(t1)* when many cells are migrating through it, followed by a late stage *(t2)* with sparse cells. Endures as the subcortical white matter.

STF5 Deep cellular layer that is prominent during the first trimester, the first sojourn zone to appear outside the germinal matrix.

Section 480

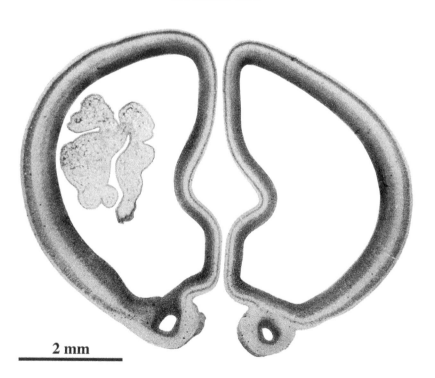

2 mm

FONT KEY:
VENTRICULAR DIVISIONS – CAPITALS
Germinal zone - Helvetica bold
Transient structure - Times bold italic
Permanent structure - Times Roman or **Bold**

Section 551

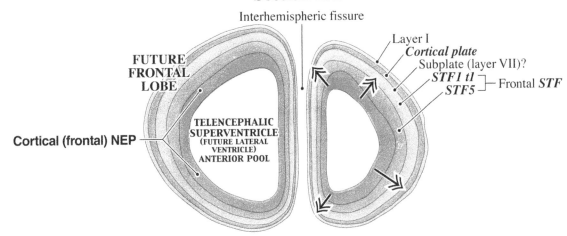

Interhemispheric fissure

Layer I
Cortical plate
Subplate (layer VII)?
STF1 tl
STF5 — Frontal *STF*

FUTURE
FRONTAL
LOBE

TELENCEPHALIC
SUPERVENTRICLE
(FUTURE LATERAL
VENTRICLE)
ANTERIOR POOL

Cortical (frontal) NEP

Section 480

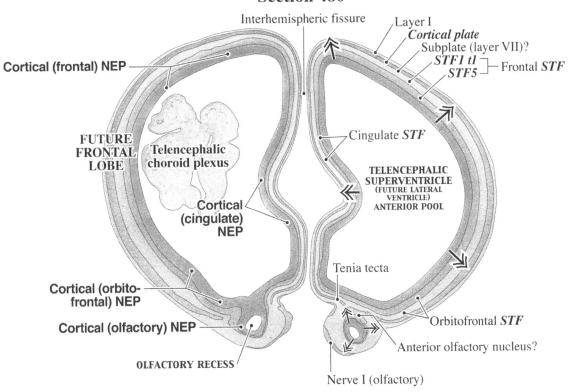

Interhemispheric fissure

Layer I
Cortical plate
Subplate (layer VII)?
STF1 tl
STF5 — Frontal *STF*

Cortical (frontal) NEP

Cingulate *STF*

FUTURE
FRONTAL
LOBE

Telencephalic
choroid plexus

TELENCEPHALIC
SUPERVENTRICLE
(FUTURE LATERAL
VENTRICLE)
ANTERIOR POOL

Cortical
(cingulate)
NEP

Tenia tecta

Cortical (orbito-
frontal) NEP

Cortical (olfactory) NEP

OLFACTORY RECESS

Orbitofrontal *STF*

Anterior olfactory nucleus?

Nerve I (olfactory)

Cortical folding in the midline
is a shrinkage artifact.

Arrows indicate the
presumed *direction of
neuron migration* from
neuroepithelial sources.

PLATE 2A
CR 42 mm,
GW 10.6, M841
Frontal/
Horizontal

Section 440

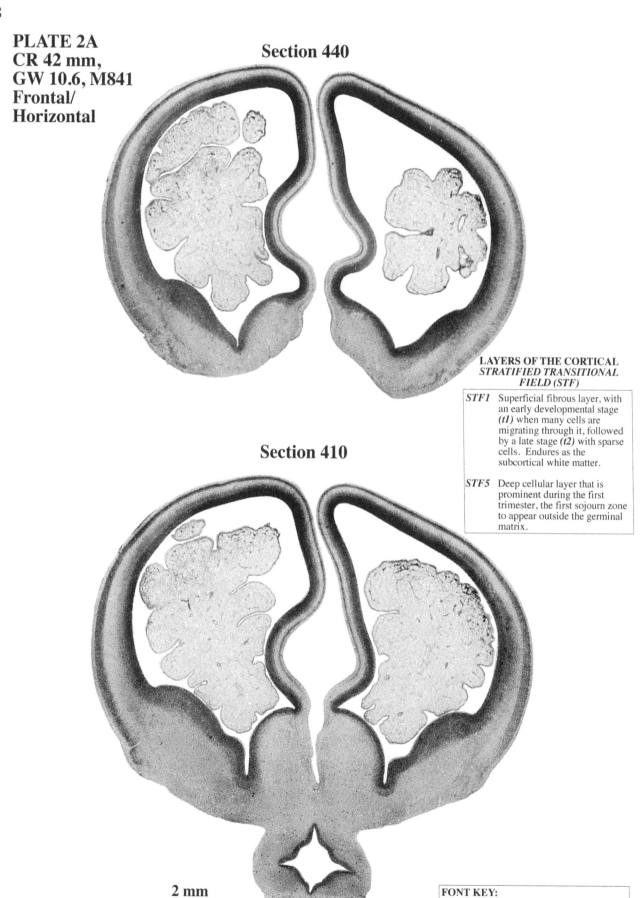

Section 410

LAYERS OF THE CORTICAL
STRATIFIED TRANSITIONAL
FIELD (STF)

STF1 Superficial fibrous layer, with an early developmental stage *(t1)* when many cells are migrating through it, followed by a late stage *(t2)* with sparse cells. Endures as the subcortical white matter.

STF5 Deep cellular layer that is prominent during the first trimester, the first sojourn zone to appear outside the germinal matrix.

2 mm

FONT KEY:
VENTRICULAR DIVISIONS – CAPITALS
Germinal zone - Helvetica bold
Transient structure - Times bold italic
Permanent structure - Times Roman or **Bold**

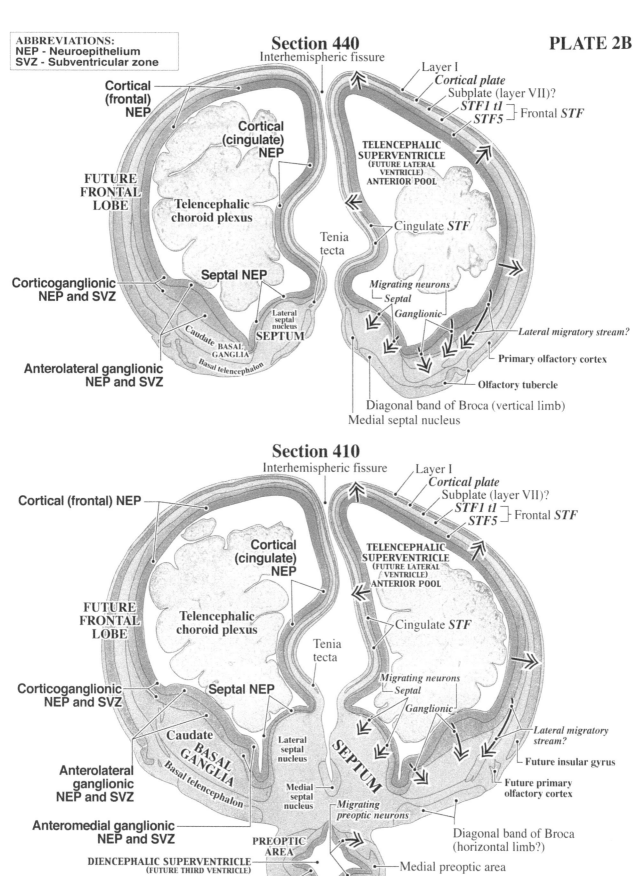

ABBREVIATIONS:
NEP - Neuroepithelium
SVZ - Subventricular zone

Section 440

PLATE 2B

Interhemispheric fissure

Layer I
Cortical plate
Subplate (layer VII)?
STF1 tl
STF5 ⎤ Frontal *STF*

Cortical (frontal) NEP

Cortical (cingulate) NEP

TELENCEPHALIC SUPERVENTRICLE
(FUTURE LATERAL VENTRICLE)
ANTERIOR POOL

FUTURE FRONTAL LOBE

Telencephalic choroid plexus

Cingulate *STF*

Tenia tecta

Migrating neurons
Septal
Ganglionic

Corticoganglionic NEP and SVZ

Septal NEP

Lateral septal nucleus

Lateral migratory stream?

Caudate
BASAL GANGLIA
Basal telencephalon

SEPTUM

Primary olfactory cortex

Anterolateral ganglionic NEP and SVZ

Olfactory tubercle

Diagonal band of Broca (vertical limb)
Medial septal nucleus

Section 410

Interhemispheric fissure

Layer I
Cortical plate
Subplate (layer VII)?
STF1 tl
STF5 ⎤ Frontal *STF*

Cortical (frontal) NEP

Cortical (cingulate) NEP

TELENCEPHALIC SUPERVENTRICLE
(FUTURE LATERAL VENTRICLE)
ANTERIOR POOL

FUTURE FRONTAL LOBE

Telencephalic choroid plexus

Tenia tecta

Cingulate *STF*

Migrating neurons
Septal
Ganglionic

Corticoganglionic NEP and SVZ

Septal NEP

Caudate
BASAL GANGLIA
Basal telencephalon

Lateral septal nucleus

Lateral migratory stream?

Future insular gyrus

Anterolateral ganglionic NEP and SVZ

Medial septal nucleus

Future primary olfactory cortex

Anteromedial ganglionic NEP and SVZ

Migrating preoptic neurons

Diagonal band of Broca (horizontal limb?)

PREOPTIC AREA

SEPTUM

Medial preoptic area

DIENCEPHALIC SUPERVENTRICLE
(FUTURE THIRD VENTRICLE)

Preoptic NEP

Optic chiasm

Suprachiasmatic nucleus

Cortical folding in the midline is a shrinkage artifact.

Arrows indicate the presumed *direction of neuron migration* from neuroepithelial sources.

**PLATE 3A
CR 42 mm,
GW 10.6, M841
Frontal/Horizontal
Section 375**

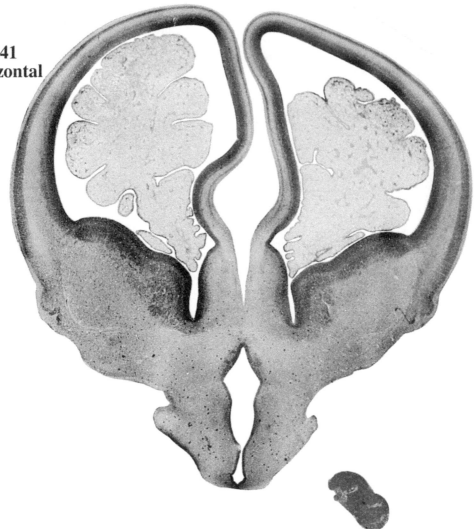

2 mm

**LAYERS OF THE CORTICAL
*STRATIFIED TRANSITIONAL FIELD (STF)***

STF1   Superficial fibrous layer with an early
developmental stage *(t1)* when many
cells are migrating through it, followed
by a late stage *(t2)* with sparse cells.
Endures as the subcortical white matter.

STF5   Deep cellular layer that is prominent
during the first trimester, the first sojourn
zone to appear outside the germinal
matrix.

**FONT KEY:
VENTRICULAR DIVISIONS – CAPITALS
Germinal zone - Helvetica bold**
Transient structure - Times bold italic
Permanent structure - Times Roman or **Bold**

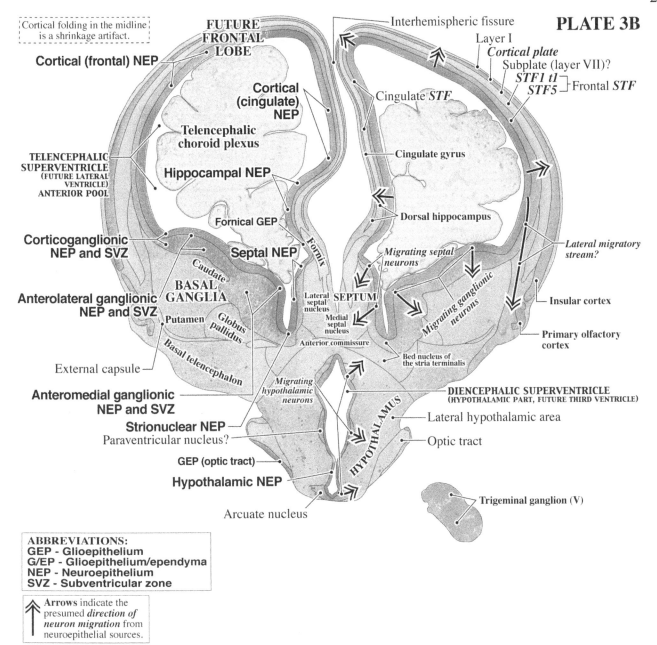

Cortical folding in the midline is a shrinkage artifact.

PLATE 3B

FUTURE FRONTAL LOBE

Interhemispheric fissure

Layer I
Cortical plate
Subplate (layer VII)?
STF1 t1
STF5 — Frontal STF

Cortical (frontal) NEP

Cortical (cingulate) NEP

Cingulate STF

Telencephalic choroid plexus

Cingulate gyrus

Hippocampal NEP

TELENCEPHALIC SUPERVENTRICLE (FUTURE LATERAL VENTRICLE) ANTERIOR POOL

Dorsal hippocampus

Fornical GEP

Corticoganglionic NEP and SVZ

Septal NEP

Caudate

BASAL GANGLIA

Fornix

Migrating septal neurons

Lateral migratory stream?

Anterolateral ganglionic NEP and SVZ

Putamen

Globus pallidus

Lateral septal nucleus

SEPTUM

Medial septal nucleus

Migrating ganglionic neurons

Insular cortex

Basal telencephalon

Anterior commissure

Primary olfactory cortex

External capsule

Bed nucleus of the stria terminalis

Anteromedial ganglionic NEP and SVZ

Migrating hypothalamic neurons

DIENCEPHALIC SUPERVENTRICLE (HYPOTHALAMIC PART, FUTURE THIRD VENTRICLE)

Strionuclear NEP

Paraventricular nucleus?

HYPOTHALAMUS

Lateral hypothalamic area

GEP (optic tract)

Optic tract

Hypothalamic NEP

Arcuate nucleus

Trigeminal ganglion (V)

ABBREVIATIONS:
GEP - Glioepithelium
G/EP - Glioepithelium/ependyma
NEP - Neuroepithelium
SVZ - Subventricular zone

Arrows indicate the presumed direction of neuron migration from neuroepithelial sources.

SPINAL CORD

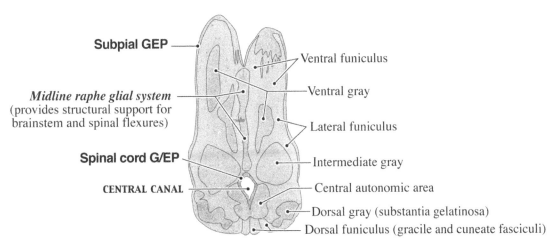

Subpial GEP

Ventral funiculus

Ventral gray

Midline raphe glial system
(provides structural support for brainstem and spinal flexures)

Lateral funiculus

Spinal cord G/EP

Intermediate gray

CENTRAL CANAL

Central autonomic area

Dorsal gray (substantia gelatinosa)

Dorsal funiculus (gracile and cuneate fasciculi)

PLATE 4A
CR 42 mm,
GW 10.6, M841
Frontal/Horizontal
Section 360

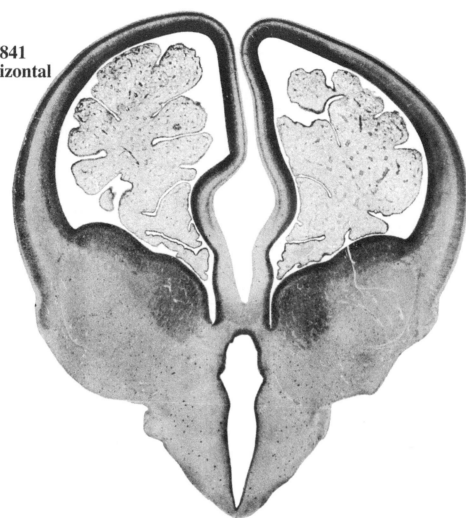

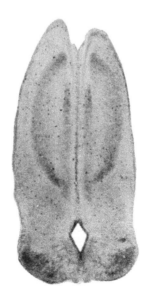

LAYERS OF THE CORTICAL
STRATIFIED TRANSITIONAL FIELD (STF)

STF1 Superficial fibrous layer with an early
developmental stage *(t1)* when many
cells are migrating through it, followed
by a late stage *(t2)* with sparse cells.
Endures as the subcortical white matter.

STF5 Deep cellular layer that is prominent
during the first trimester, the first sojourn
zone to appear outside the germinal
matrix.

FONT KEY:
VENTRICULAR DIVISIONS – CAPITALS
Germinal zone - Helvetica bold
Transient structure - Times bold italic
Permanent structure - Times Roman or **Bold**

2 mm

PLATE 4B

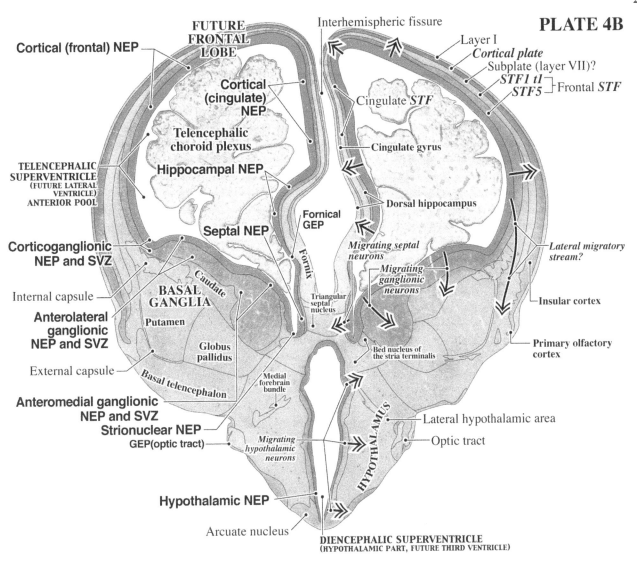

Cortical (frontal) NEP

FUTURE FRONTAL LOBE

Interhemispheric fissure

Layer I
Cortical plate
Subplate (layer VII)?
STF1 tl
STF5 ⎤ Frontal *STF*

Cortical (cingulate) NEP

Cingulate *STF*

Telencephalic choroid plexus

Cingulate gyrus

TELENCEPHALIC SUPERVENTRICLE (FUTURE LATERAL VENTRICLE) ANTERIOR POOL

Hippocampal NEP

Dorsal hippocampus

Fornical GEP

Septal NEP

Migrating septal neurons

Migrating ganglionic neurons

Lateral migratory stream?

Corticoganglionic NEP and SVZ

Fornix

Internal capsule

Caudate

BASAL GANGLIA

Triangular septal nucleus

Insular cortex

Anterolateral ganglionic NEP and SVZ

Putamen

Bed nucleus of the stria terminalis

Primary olfactory cortex

Globus pallidus

External capsule

Basal telencephalon

Medial forebrain bundle

Anteromedial ganglionic NEP and SVZ

Strionuclear NEP

GEP(optic tract)

Migrating hypothalamic neurons

HYPOTHALAMUS

Lateral hypothalamic area

Optic tract

Hypothalamic NEP

Arcuate nucleus

DIENCEPHALIC SUPERVENTRICLE (HYPOTHALAMIC PART, FUTURE THIRD VENTRICLE)

ABBREVIATIONS:
GEP - Glioepithelium
G/EP - Glioepithelium/ependyma
NEP - Neuroepithelium
SVZ - Subventricular zone

Arrows indicate the presumed *direction of neuron migration* from neuroepithelial sources.

MEDULLA

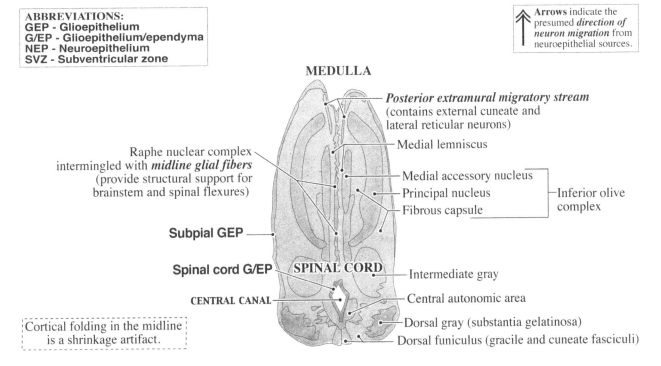

Posterior extramural migratory stream (contains external cuneate and lateral reticular neurons)

Raphe nuclear complex intermingled with *midline glial fibers* (provide structural support for brainstem and spinal flexures)

Medial lemniscus

Medial accessory nucleus
Principal nucleus
Fibrous capsule

Inferior olive complex

Subpial GEP

Spinal cord G/EP

SPINAL CORD

Intermediate gray

CENTRAL CANAL

Central autonomic area

Dorsal gray (substantia gelatinosa)

Cortical folding in the midline is a shrinkage artifact.

Dorsal funiculus (gracile and cuneate fasciculi)

PLATE 5A
CR 42 mm, GW 10.6,
M841
Frontal/
Horizontal
Section 340

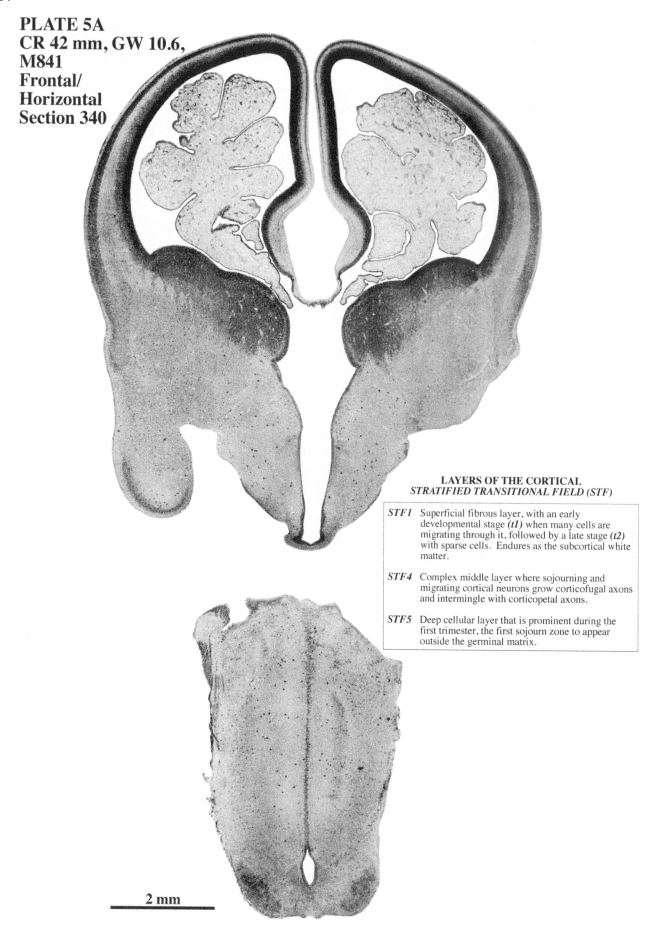

LAYERS OF THE CORTICAL
STRATIFIED TRANSITIONAL FIELD (STF)

STF1 Superficial fibrous layer, with an early
developmental stage *(t1)* when many cells are
migrating through it, followed by a late stage *(t2)*
with sparse cells. Endures as the subcortical white
matter.

STF4 Complex middle layer where sojourning and
migrating cortical neurons grow corticofugal axons
and intermingle with corticopetal axons.

STF5 Deep cellular layer that is prominent during the
first trimester, the first sojourn zone to appear
outside the germinal matrix.

2 mm

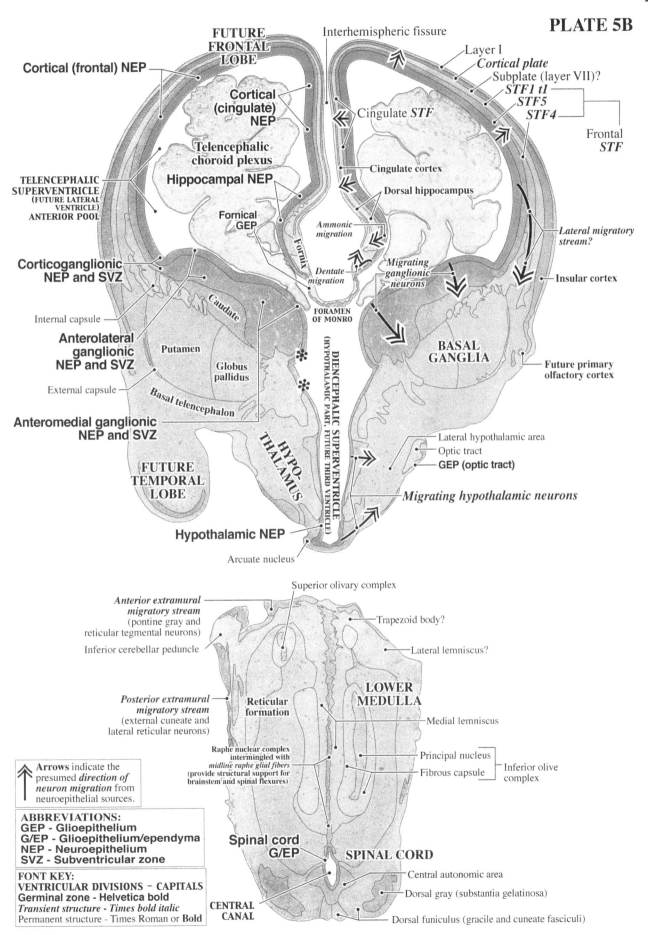

FUTURE FRONTAL LOBE

Interhemispheric fissure

Layer I
Cortical plate
Subplate (layer VII)?
STF1 t1
STF5
STF4

Frontal *STF*

Cortical (frontal) NEP

Cortical (cingulate) NEP

Cingulate *STF*

Telencephalic choroid plexus

Cingulate cortex

Hippocampal NEP

Dorsal hippocampus

TELENCEPHALIC SUPERVENTRICLE (FUTURE LATERAL VENTRICLE) ANTERIOR POOL

Fornical GEP

Fornix

Ammonic migration

Lateral migratory stream?

Corticoganglionic NEP and SVZ

Dentate migration

Migrating ganglionic neurons

Insular cortex

Internal capsule

Caudate

FORAMEN OF MONRO

Anterolateral ganglionic NEP and SVZ

Putamen

DIENCEPHALIC SUPERVENTRICLE (HYPOTHALAMIC PART, FUTURE THIRD VENTRICLE)

BASAL GANGLIA

Globus pallidus

External capsule

Basal telencephalon

Future primary olfactory cortex

Anteromedial ganglionic NEP and SVZ

FUTURE TEMPORAL LOBE

HYPO-THALAMUS

Lateral hypothalamic area
Optic tract
GEP (optic tract)

Migrating hypothalamic neurons

Hypothalamic NEP

Arcuate nucleus

Superior olivary complex

Anterior extramural migratory stream
(pontine gray and reticular tegmental neurons)

Trapezoid body?

Inferior cerebellar peduncle

Lateral lemniscus?

Posterior extramural migratory stream
(external cuneate and lateral reticular neurons)

Reticular formation

LOWER MEDULLA

Medial lemniscus

Raphe nuclear complex intermingled with *midline raphe glial fibers* (provide structural support for brainstem and spinal flexures)

Principal nucleus

Fibrous capsule

Inferior olive complex

Spinal cord G/EP

SPINAL CORD

Central autonomic area

Dorsal gray (substantia gelatinosa)

CENTRAL CANAL

Dorsal funiculus (gracile and cuneate fasciculi)

Arrows indicate the presumed *direction of neuron migration* from neuroepithelial sources.

ABBREVIATIONS:
GEP - Glioepithelium
G/EP - Glioepithelium/ependyma
NEP - Neuroepithelium
SVZ - Subventricular zone

FONT KEY:
VENTRICULAR DIVISIONS – CAPITALS
Germinal zone - Helvetica bold
Transient structure - Times bold italic
Permanent structure - Times Roman or **Bold**

PLATE 6A
CR 42 mm, GW 10.6,
M841
Frontal/Horizontal
Section 320

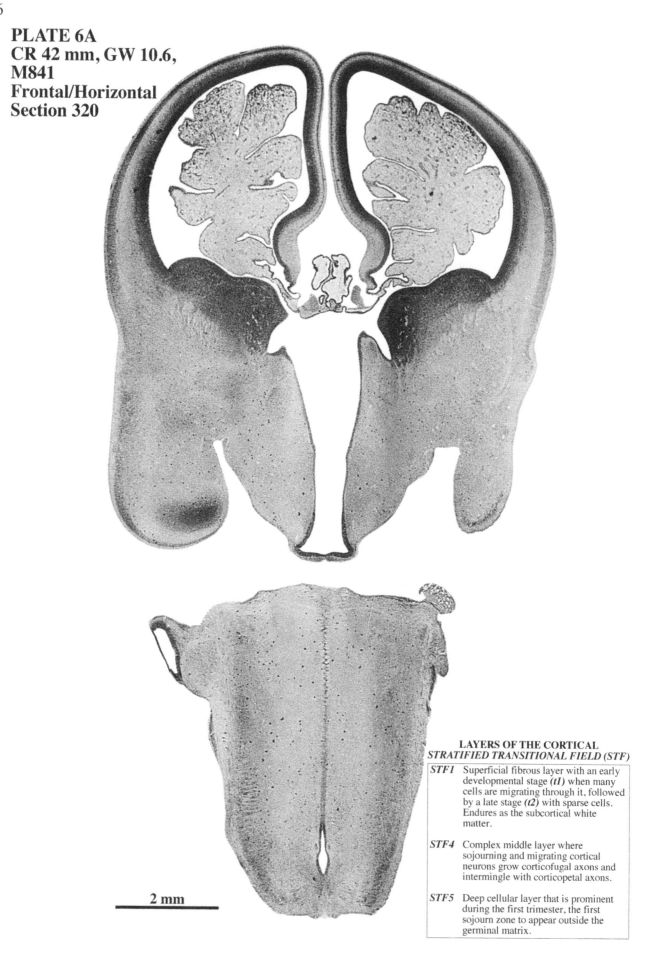

2 mm

LAYERS OF THE CORTICAL
STRATIFIED TRANSITIONAL FIELD (STF)

STF1	Superficial fibrous layer with an early developmental stage *(t1)* when many cells are migrating through it, followed by a late stage *(t2)* with sparse cells. Endures as the subcortical white matter.
STF4	Complex middle layer where sojourning and migrating cortical neurons grow corticofugal axons and intermingle with corticopetal axons.
STF5	Deep cellular layer that is prominent during the first trimester, the first sojourn zone to appear outside the germinal matrix.

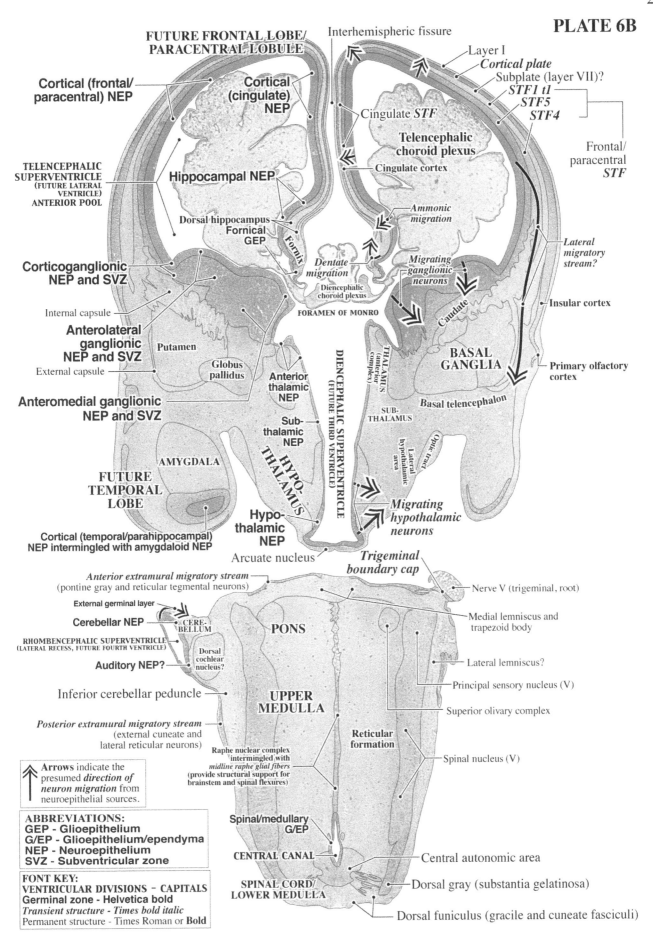

FUTURE FRONTAL LOBE/
PARACENTRAL LOBULE

Interhemispheric fissure

Layer I
Cortical plate
Subplate (layer VII)?
STF1 t1
STF5
STF4

Frontal/
paracentral
STF

Cortical (frontal/
paracentral) NEP

Cortical
(cingulate)
NEP

Cingulate *STF*

Telencephalic
choroid plexus

Cingulate cortex

TELENCEPHALIC
SUPERVENTRICLE
(FUTURE LATERAL
VENTRICLE)
ANTERIOR POOL

Hippocampal NEP

Dorsal hippocampus
Fornical
GEP

Fornix

*Ammonic
migration*

*Dentate
migration*

Diencephalic
choroid plexus

*Migrating
ganglionic
neurons*

*Lateral
migratory
stream?*

Corticoganglionic
NEP and SVZ

Internal capsule

FORAMEN OF MONRO

Caudate

Insular cortex

Anterolateral
ganglionic
NEP and SVZ

External capsule

Putamen

Globus
pallidus

Anterior
thalamic
NEP

DIENCEPHALIC SUPERVENTRICLE
(FUTURE THIRD VENTRICLE)

THALAMUS
(anterior
complex)

BASAL
GANGLIA

Basal telencephalon

SUB-
THALAMUS

Primary olfactory
cortex

Anteromedial ganglionic
NEP and SVZ

Sub-
thalamic
NEP

HYPO-
THALAMUS

Lateral
hypothalamic
area

Optic tract

AMYGDALA

FUTURE
TEMPORAL
LOBE

Hypo-
thalamic
NEP

*Migrating
hypothalamic
neurons*

Cortical (temporal/parahippocampal)
NEP intermingled with amygdaloid NEP

Arcuate nucleus

*Trigeminal
boundary cap*

Nerve V (trigeminal, root)

Anterior extramural migratory stream
(pontine gray and reticular tegmental neurons)

External germinal layer

Cerebellar NEP

CERE-
BELLUM

PONS

Medial lemniscus and
trapezoid body

RHOMBENCEPHALIC SUPERVENTRICLE
(LATERAL RECESS, FUTURE FOURTH VENTRICLE)

Dorsal
cochlear
nucleus?

Lateral lemniscus?

Auditory NEP?

Principal sensory nucleus (V)

Inferior cerebellar peduncle

UPPER
MEDULLA

Superior olivary complex

Posterior extramural migratory stream
(external cuneate and
lateral reticular neurons)

Raphe nuclear complex
intermingled with
midline raphe glial fibers
(provide structural support for
brainstem and spinal flexures)

Reticular
formation

Spinal nucleus (V)

Arrows indicate the
presumed *direction of
neuron migration* from
neuroepithelial sources.

Spinal/medullary
G/EP

ABBREVIATIONS:
GEP - Glioepithelium
G/EP - Glioepithelium/ependyma
NEP - Neuroepithelium
SVZ - Subventricular zone

CENTRAL CANAL

Central autonomic area

FONT KEY:
VENTRICULAR DIVISIONS – CAPITALS
Germinal zone - Helvetica bold
Transient structure - Times bold italic
Permanent structure - Times Roman or **Bold**

SPINAL CORD/
LOWER MEDULLA

Dorsal gray (substantia gelatinosa)

Dorsal funiculus (gracile and cuneate fasciculi)

PLATE 7A
CR 42 mm, GW 10.6, M841
Frontal/Horizontal
Section 294

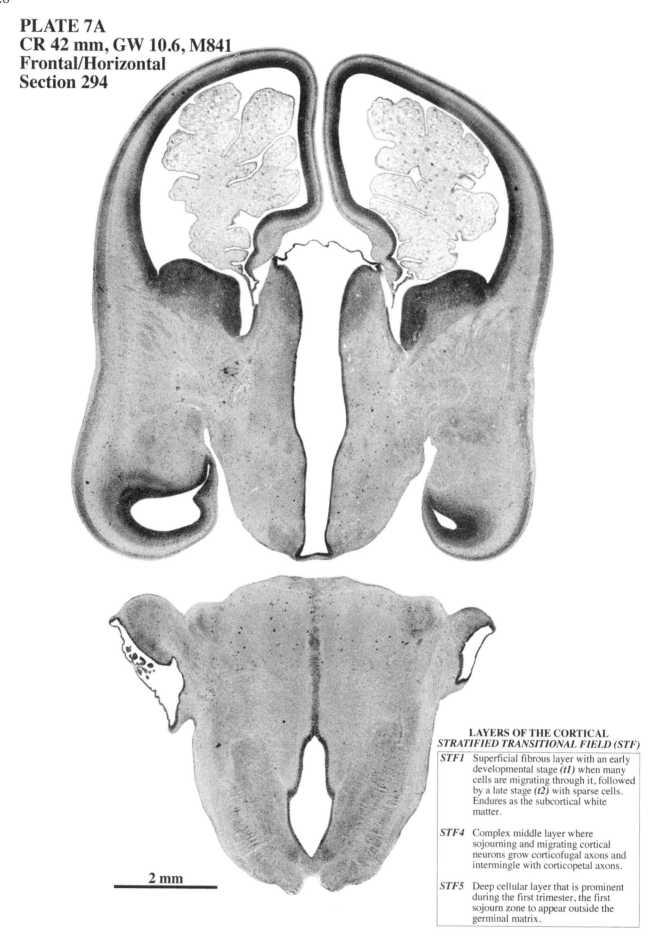

2 mm

LAYERS OF THE CORTICAL
STRATIFIED TRANSITIONAL FIELD (STF)

STF1	Superficial fibrous layer with an early developmental stage *(t1)* when many cells are migrating through it, followed by a late stage *(t2)* with sparse cells. Endures as the subcortical white matter.
STF4	Complex middle layer where sojourning and migrating cortical neurons grow corticofugal axons and intermingle with corticopetal axons.
STF5	Deep cellular layer that is prominent during the first trimester, the first sojourn zone to appear outside the germinal matrix.

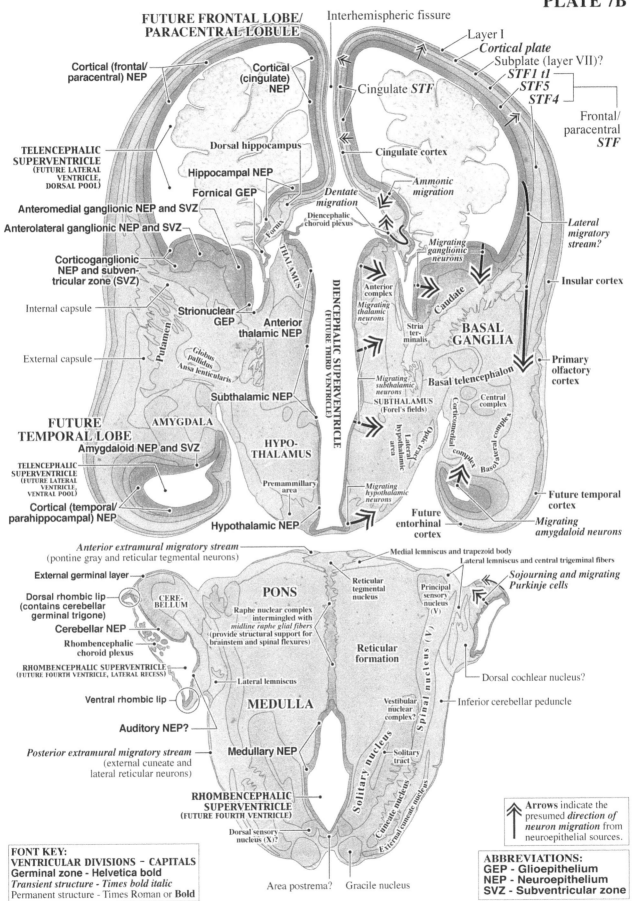

FUTURE FRONTAL LOBE/
PARACENTRAL LOBULE

Interhemispheric fissure

Layer I
Cortical plate
Subplate (layer VII)?
STF1 tl
STF5
STF4

Frontal/
paracentral
STF

Cortical (frontal/
paracentral) NEP

Cortical
(cingulate)
NEP

Cingulate *STF*

TELECEPHALIC
SUPERVENTRICLE
(FUTURE LATERAL
VENTRICLE,
DORSAL POOL)

Dorsal hippocampus

Cingulate cortex

Hippocampal NEP

Fornical GEP

*Dentate
migration*

*Ammonic
migration*

Diencephalic
choroid plexus

Anteromedial ganglionic NEP and SVZ

Anterolateral ganglionic NEP and SVZ

Fornix

*Migrating
ganglionic
neurons*

*Lateral
migratory
stream?*

Corticoganglionic
NEP and subven-
tricular zone (SVZ)

THALAMUS

Anterior
complex

Caudate

Insular cortex

Internal capsule

Strionuclear
GEP

Anterior
thalamic NEP

*Migrating
thalamic
neurons*

Stria
ter-
minalis

BASAL
GANGLIA

Putamen

Globus
pallidus

Ansa lenticularis

DIENCEPHALIC SUPERVENTRICLE (FUTURE THIRD VENTRICLE)

Primary
olfactory
cortex

External capsule

Subthalamic NEP

Basal telencephalon

Central
complex

FUTURE
TEMPORAL LOBE

AMYGDALA

*Migrating
subthalamic
neurons*

SUBTHALAMUS
(Forel's fields)

Corticomedial
complex

Basolateral complex

Amygdaloid NEP and SVZ

HYPO-
THALAMUS

Lateral
hypothalamic
area

Optic tract

TELECEPHALIC
SUPERVENTRICLE
(FUTURE LATERAL
VENTRICLE,
VENTRAL POOL)

Premammillary
area

*Migrating
hypothalamic
neurons*

Future temporal
cortex

Cortical (temporal/
parahippocampal) NEP

Hypothalamic NEP

Future
entorhinal
cortex

*Migrating
amygdaloid neurons*

Anterior extramural migratory stream
(pontine gray and reticular tegmental neurons)

Medial lemniscus and trapezoid body

Lateral lemniscus and central trigeminal fibers

External germinal layer

PONS

Reticular
tegmental
nucleus

Principal
sensory
nucleus
(V)

*Sojourning and migrating
Purkinje cells*

Dorsal rhombic lip
(contains cerebellar
germinal trigone)

CERE-
BELLUM

Cerebellar NEP

Raphe nuclear complex
intermingled with
midline raphe glial fibers
(provide structural support for
brainstem and spinal flexures)

Reticular
formation

Rhombencephalic
choroid plexus

Spinal nucleus (V)

RHOMBENCEPHALIC SUPERVENTRICLE
(FUTURE FOURTH VENTRICLE, LATERAL RECESS)

Lateral lemniscus

Dorsal cochlear nucleus?

Ventral rhombic lip

MEDULLA

Vestibular
nuclear
complex?

Inferior cerebellar peduncle

Auditory NEP?

Solitary nucleus

Solitary
tract

Posterior extramural migratory stream
(external cuneate and
lateral reticular neurons)

Medullary NEP

Cuneate nucleus

External cuneate nucleus

RHOMBENCEPHALIC
SUPERVENTRICLE
(FUTURE FOURTH VENTRICLE)

Dorsal sensory
nucleus (X)?

Area postrema?

Gracile nucleus

Arrows indicate the
presumed *direction of
neuron migration* from
neuroepithelial sources.

FONT KEY:
VENTRICULAR DIVISIONS – CAPITALS
Germinal zone - Helvetica bold
Transient structure - Times bold italic
Permanent structure - Times Roman or **Bold**

ABBREVIATIONS:
GEP - Glioepithelium
NEP - Neuroepithelium
SVZ - Subventricular zone

PLATE 8A
CR 42 mm, GW 10.6, M841
Frontal/Horizontal
Section 285

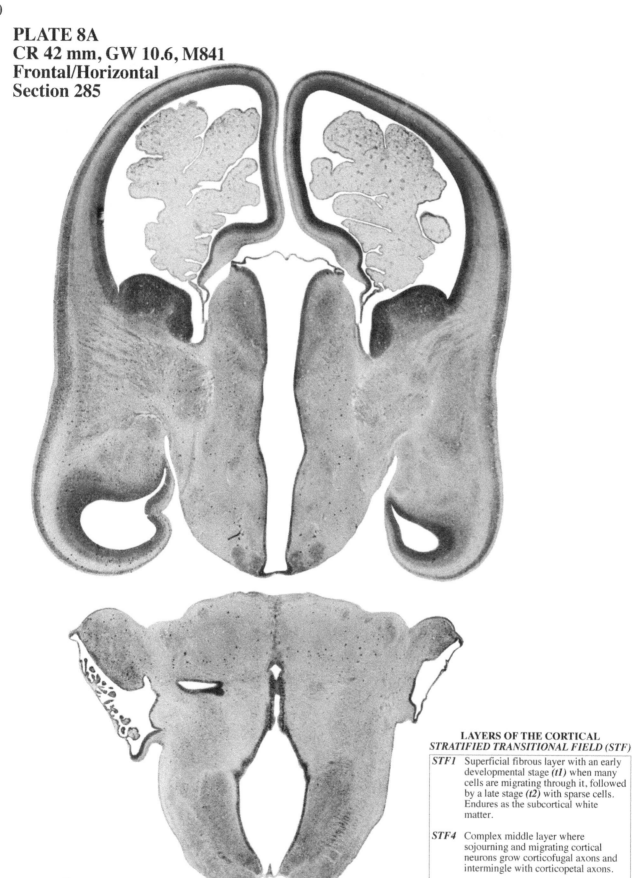

2 mm

LAYERS OF THE CORTICAL
STRATIFIED TRANSITIONAL FIELD (STF)

STF1	Superficial fibrous layer with an early developmental stage *(t1)* when many cells are migrating through it, followed by a late stage *(t2)* with sparse cells. Endures as the subcortical white matter.
STF4	Complex middle layer where sojourning and migrating cortical neurons grow corticofugal axons and intermingle with corticopetal axons.
STF5	Deep cellular layer that is prominent during the first trimester, the first sojourn zone to appear outside the germinal matrix.

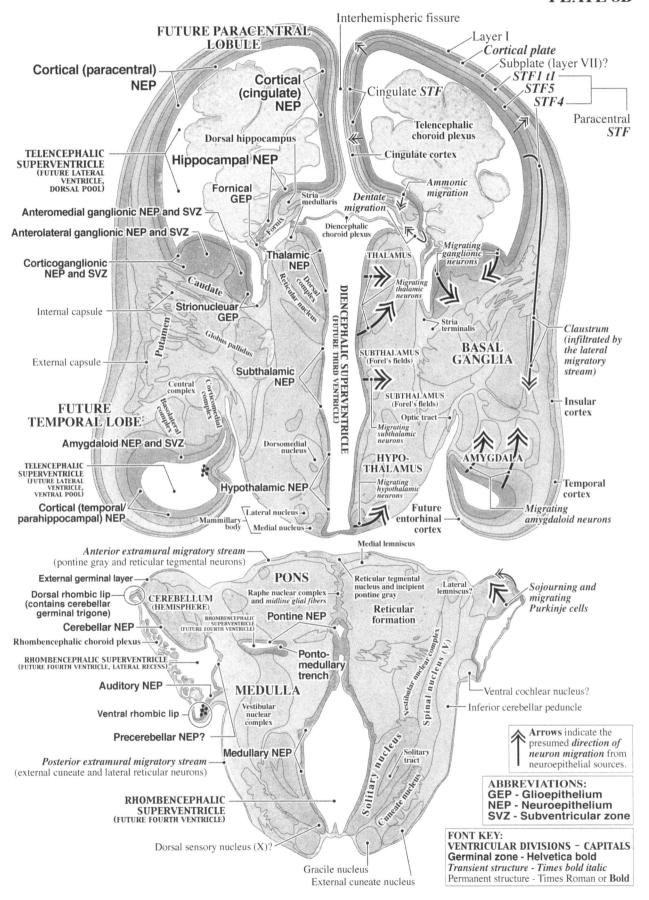

FUTURE PARACENTRAL LOBULE

Cortical (paracentral) NEP

Cortical (cingulate) NEP

TELENCEPHALIC SUPERVENTRICLE (FUTURE LATERAL VENTRICLE, DORSAL POOL)

Hippocampal NEP

Dorsal hippocampus

Fornical GEP

Anteromedial ganglionic NEP and SVZ

Anterolateral ganglionic NEP and SVZ

Corticoganglionic NEP and SVZ

Internal capsule

Strionucleuar GEP

Caudate

External capsule

FUTURE TEMPORAL LOBE

Amygdaloid NEP and SVZ

TELENCEPHALIC SUPERVENTRICLE (FUTURE LATERAL VENTRICLE, VENTRAL POOL)

Cortical (temporal/parahippocampal) NEP

Putamen

Globus pallidus

Central complex

Corticomedial complex

Basolateral complex

Subthalamic NEP

Dorsomedial nucleus

Hypothalamic NEP

Mammillary body

Lateral nucleus

Medial nucleus

Stria medullaris

Fornix

Thalamic NEP

Dorsal complex

Reticular nucleus

Interhemispheric fissure

Layer I

Cortical plate

Subplate (layer VII)?

STF1 t1
STF5
STF4

Paracentral STF

Cingulate STF

Telencephalic choroid plexus

Cingulate cortex

Ammonic migration

Dentate migration

Diencephalic choroid plexus

DIENCEPHALIC SUPERVENTRICLE (FUTURE THIRD VENTRICLE)

THALAMUS

Migrating thalamic neurons

SUBTHALAMUS (Forel's fields)

SUBTHALAMUS (Forel's fields)

Optic tract

Migrating subthalamic neurons

HYPO-THALAMUS

Migrating hypothalamic neurons

Future entorhinal cortex

Migrating ganglionic neurons

Stria terminalis

BASAL GANGLIA

Claustrum (infiltrated by the lateral migratory stream)

Insular cortex

AMYGDALA

Temporal cortex

Migrating amygdaloid neurons

Medial lemniscus

Anterior extramural migratory stream (pontine gray and reticular tegmental neurons)

External germinal layer

Dorsal rhombic lip (contains cerebellar germinal trigone)

Cerebellar NEP

Rhombencephalic choroid plexus

RHOMBENCEPHALIC SUPERVENTRICLE (FUTURE FOURTH VENTRICLE, LATERAL RECESS)

Auditory NEP

Ventral rhombic lip

Precerebellar NEP?

Medullary NEP

Posterior extramural migratory stream (external cuneate and lateral reticular neurons)

RHOMBENCEPHALIC SUPERVENTRICLE (FUTURE FOURTH VENTRICLE)

PONS

CEREBELLUM (HEMISPHERE)

RHOMBENCEPHALIC SUPERVENTRICLE (FUTURE FOURTH VENTRICLE)

Raphe nuclear complex and midline glial fibers

Pontine NEP

Ponto-medullary trench

MEDULLA

Vestibular nuclear complex

Reticular tegmental nucleus and incipient pontine gray

Reticular formation

Vestibular nuclear complex (V)

Spinal nucleus (V)

Lateral lemniscus?

Solitary tract

Solitary nucleus

Cuneate nucleus

Dorsal sensory nucleus (X)?

Gracile nucleus

External cuneate nucleus

Sojourning and migrating Purkinje cells

Ventral cochlear nucleus?

Inferior cerebellar peduncle

Arrows indicate the presumed *direction of neuron migration* from neuroepithelial sources.

ABBREVIATIONS:
GEP - Glioepithelium
NEP - Neuroepithelium
SVZ - Subventricular zone

FONT KEY:
VENTRICULAR DIVISIONS – CAPITALS
Germinal zone - Helvetica bold
Transient structure - Times bold italic
Permanent structure - Times Roman or Bold

PLATE 9A
CR 42 mm,
GW 10.6, M841
Frontal/Horizontal
Section 270

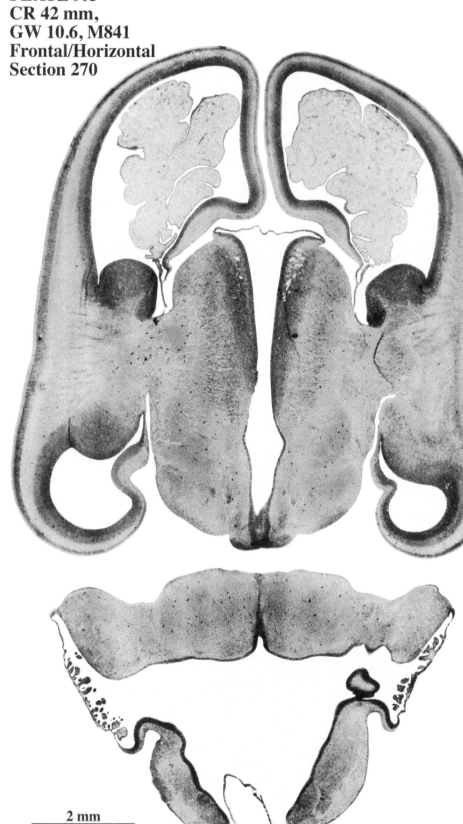

2 mm

LAYERS OF THE
CORTICAL *STRATIFIED*
TRANSITIONAL FIELD (STF)

STF1 Superficial fibrous layer with
an early developmental stage
(t1) when many cells are
migrating through it,
followed by a late stage *(t2)*
with sparse cells. Endures as
the subcortical white matter.

STF4 Complex middle layer where
sojourning and migrating
cortical neurons grow
corticofugal axons and
intermingle with corticopetal
axons.

STF5 Deep cellular layer that is
prominent during the first
trimester, the first sojourn
zone to appear outside the
germinal matrix.

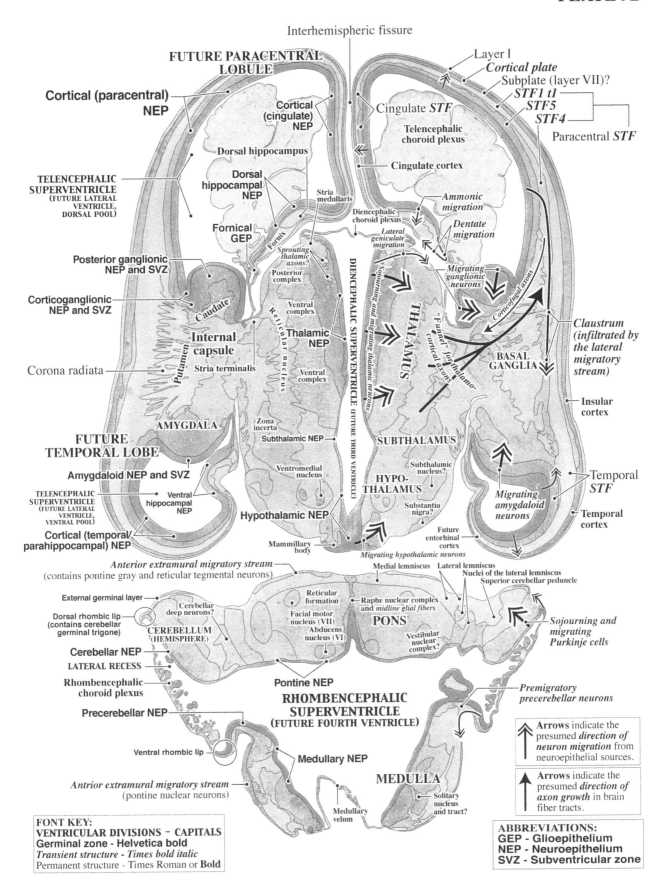

Interhemispheric fissure

FUTURE PARACENTRAL LOBULE

Layer I
Cortical plate
Subplate (layer VII)?
STF1 t1
STF5
STF4
Paracentral *STF*

Cortical (paracentral) NEP

Cortical (cingulate) NEP

Cingulate *STF*

Telencephalic choroid plexus

Dorsal hippocampus

Cingulate cortex

Dorsal hippocampal NEP

Stria medullaris

Ammonic migration

TELENCEPHALIC SUPERVENTRICLE (FUTURE LATERAL VENTRICLE, DORSAL POOL)

Diencephalic choroid plexus

Dentate migration

Fornical GEP

Fornix

Lateral geniculate migration

Sprouting thalamic axons?

Posterior complex

Migrating ganglionic neurons

Posterior ganglionic NEP and SVZ

Corticoganglionic NEP and SVZ

Caudate

Ventral complex

Thalamic NEP

Reticular nucleus

Sojourning and migrating thalamic neurons

THALAMUS

"Fuimen" for thalamo-cortical axons

Corticofugal axons

Claustrum (infiltrated by the lateral migratory stream)

Internal capsule

Putamen

Corona radiata

Stria terminalis

DIENCEPHALIC SUPERVENTRICLE (FUTURE THIRD VENTRICLE)

Ventral complex

BASAL GANGLIA

Insular cortex

AMYGDALA

Zona incerta

Subthalamic NEP

SUBTHALAMUS

FUTURE TEMPORAL LOBE

Ventromedial nucleus

Subthalamic nucleus?

HYPO-THALAMUS

Temporal *STF*

Amygdaloid NEP and SVZ

Ventral hippocampal NEP

TELENCEPHALIC SUPERVENTRICLE (FUTURE LATERAL VENTRICLE, VENTRAL POOL)

Hypothalamic NEP

Substantia nigra?

Migrating amygdaloid neurons

Temporal cortex

Cortical (temporal/ parahippocampal) NEP

Mammillary body

Future entorhinal cortex

Migrating hypothalamic neurons

Anterior extramural migratory stream
(contains pontine gray and reticular tegmental neurons)

Medial lemniscus

Lateral lemniscus

Nuclei of the lateral lemniscus

Superior cerebellar peduncle

External germinal layer

Cerebellar deep neurons?

Reticular formation

Raphe nuclear complex and *midline glial fibers*

Dorsal rhombic lip (contains cerebellar germinal trigone)

Facial motor nucleus (VII)

PONS

Sojourning and migrating Purkinje cells

CEREBELLUM (HEMISPHERE)

Abducens nucleus (VI)

Vestibular nuclear complex?

Cerebellar NEP

LATERAL RECESS

Rhombencephalic choroid plexus

Premigratory precerebellar neurons

Pontine NEP

Precerebellar NEP

RHOMBENCEPHALIC SUPERVENTRICLE (FUTURE FOURTH VENTRICLE)

Ventral rhombic lip

Medullary NEP

↑ **Arrows** indicate the presumed *direction of neuron migration* from neuroepithelial sources.

↑ **Arrows** indicate the presumed *direction of axon growth* in brain fiber tracts.

Antrior extramural migratory stream
(pontine nuclear neurons)

MEDULLA

Solitary nucleus and tract?

Medullary velum

FONT KEY:
VENTRICULAR DIVISIONS – CAPITALS
Germinal zone - Helvetica bold
Transient structure - Times bold italic
Permanent structure - Times Roman or **Bold**

ABBREVIATIONS:
GEP - Glioepithelium
NEP - Neuroepithelium
SVZ - Subventricular zone

PLATE 10A
CR 42 mm,
GW 10.6, M841
Frontal/Horizontal
Section 255

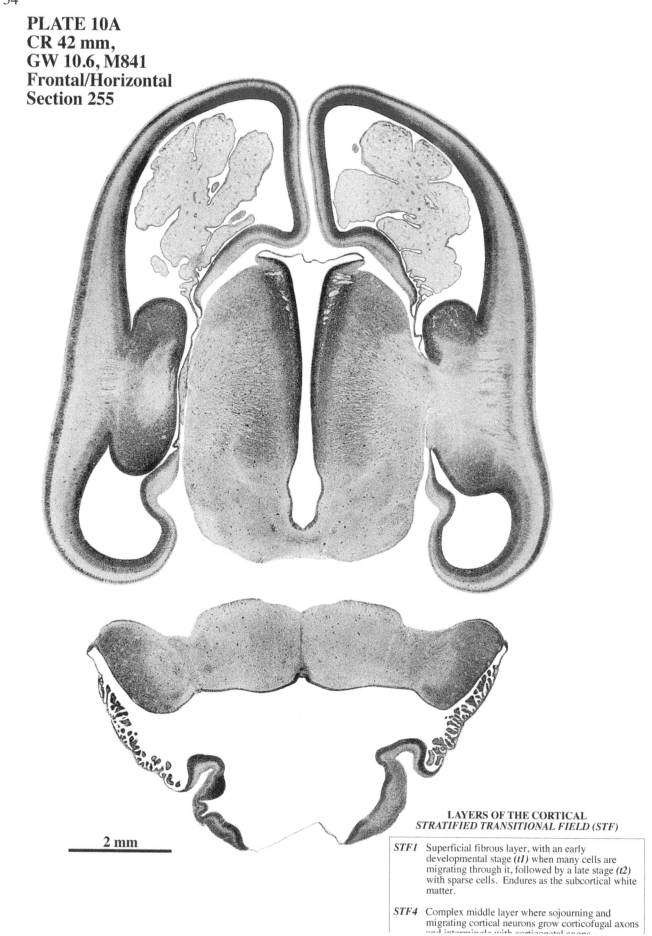

2 mm

LAYERS OF THE CORTICAL
STRATIFIED TRANSITIONAL FIELD (STF)

STF1 Superficial fibrous layer, with an early
developmental stage *(t1)* when many cells are
migrating through it, followed by a late stage *(t2)*
with sparse cells. Endures as the subcortical white
matter.

STF4 Complex middle layer where sojourning and
migrating cortical neurons grow corticofugal axons
and intermingle with corticopetal axons

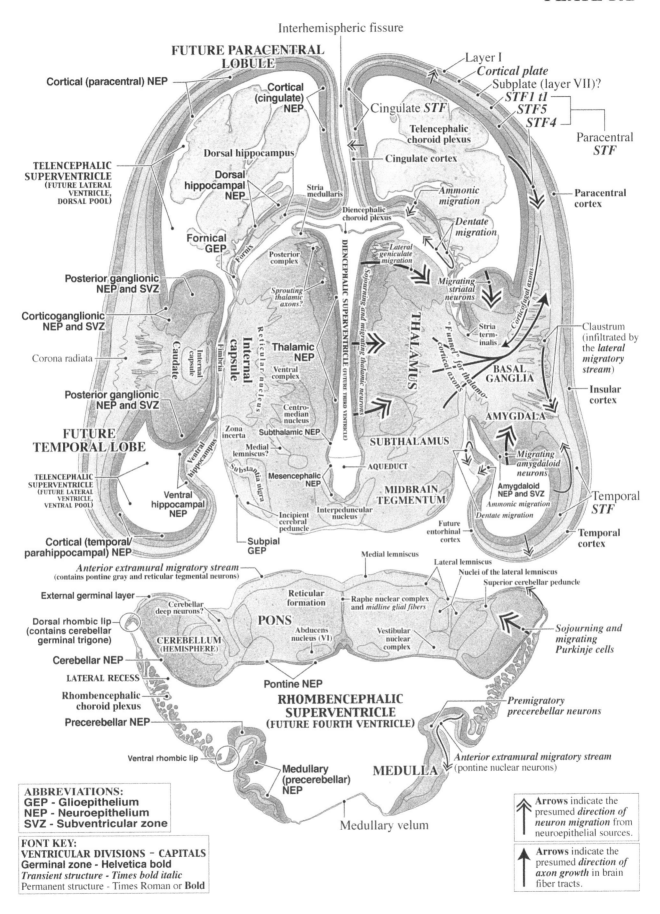

Interhemispheric fissure

FUTURE PARACENTRAL
LOBULE

Layer I
Cortical plate
Subplate (layer VII)?
STF1 tl
STF5
STF4
Paracentral
STF

Cortical (paracentral) NEP

Cortical
(cingulate)
NEP

Cingulate *STF*

Telencephalic
choroid plexus

Cingulate cortex

Dorsal hippocampus

Dorsal
hippocampal
NEP

Stria
medullaris

Diencephalic
choroid plexus

*Ammonic
migration*

Paracentral
cortex

TELENCEPHALIC
SUPERVENTRICLE
(FUTURE LATERAL
VENTRICLE,
DORSAL POOL)

*Dentate
migration*

Fornical
GEP

Fornix

Posterior
complex

*Lateral
geniculate
migration*

Posterior ganglionic
NEP and SVZ

*Sprouting
thalamic
axons?*

*Migrating
striatal
neurons*

Claustrum
(infiltrated by
the *lateral
migratory
stream*)

Corticoganglionic
NEP and SVZ

Corona radiata

Caudate

Internal
capsule

Fimbria

Reticular nucleus

Internal
capsule

Thalamic
NEP

Sojourning and migrating thalamic neurons

DIENCEPHALIC SUPERVENTRICLE (FUTURE THIRD VENTRICLE)

THALAMUS

Stria
terminalis

Corticofugal axons

*"Funnel" for thalamo-
cortical axons*

BASAL
GANGLIA

Insular
cortex

Ventral
complex

Posterior ganglionic
NEP and SVZ

Centro-
median
nucleus

AMYGDALA

FUTURE
TEMPORAL LOBE

Zona
incerta

Subthalamic NEP

*Migrating
amygdaloid
neurons*

*Medial
lemniscus?*

SUBTHALAMUS

TELENCEPHALIC
SUPERVENTRICLE
(FUTURE LATERAL
VENTRICLE,
VENTRAL POOL)

Substantia nigra

Ventral hippocampus

Mesencephalic
NEP

AQUEDUCT

Amygdaloid
NEP and SVZ

Ammonic migration

Temporal
STF

Ventral
hippocampal
NEP

MIDBRAIN
TEGMENTUM

Dentate migration

*Incipient
cerebral
peduncle*

Interpeduncular
nucleus

Future
entorhinal
cortex

Temporal
cortex

Cortical (temporal/
parahippocampal) NEP

Subpial
GEP

Anterior extramural migratory stream
(contains pontine gray and reticular tegmental neurons)

Medial lemniscus

Lateral lemniscus

Nuclei of the lateral lemniscus

Superior cerebellar peduncle

External germinal layer

*Cerebellar
deep neurons?*

Reticular
formation

Raphe nuclear complex
and *midline glial fibers*

Dorsal rhombic lip
(contains cerebellar
germinal trigone)

PONS

Abducens
nucleus (VI)

Vestibular
nuclear
complex

*Sojourning and
migrating
Purkinje cells*

Cerebellar NEP

CEREBELLUM
(HEMISPHERE)

LATERAL RECESS

Rhombencephalic
choroid plexus

Pontine NEP

RHOMBENCEPHALIC
SUPERVENTRICLE
(FUTURE FOURTH VENTRICLE)

*Premigratory
precerebellar neurons*

Precerebellar NEP

Ventral rhombic lip

Medullary
(precerebellar)
NEP

MEDULLA

Anterior extramural migratory stream
(pontine nuclear neurons)

Medullary velum

ABBREVIATIONS:
GEP - Glioepithelium
NEP - Neuroepithelium
SVZ - Subventricular zone

FONT KEY:
VENTRICULAR DIVISIONS – CAPITALS
Germinal zone - Helvetica bold
Transient structure - Times bold italic
Permanent structure - Times Roman or **Bold**

Arrows indicate the
presumed *direction of
neuron migration* from
neuroepithelial sources.

Arrows indicate the
presumed *direction of
axon growth* in brain
fiber tracts.

PLATE 11A
CR 42 mm,
GW 10.6, M841
Frontal/Horizontal
Section 235

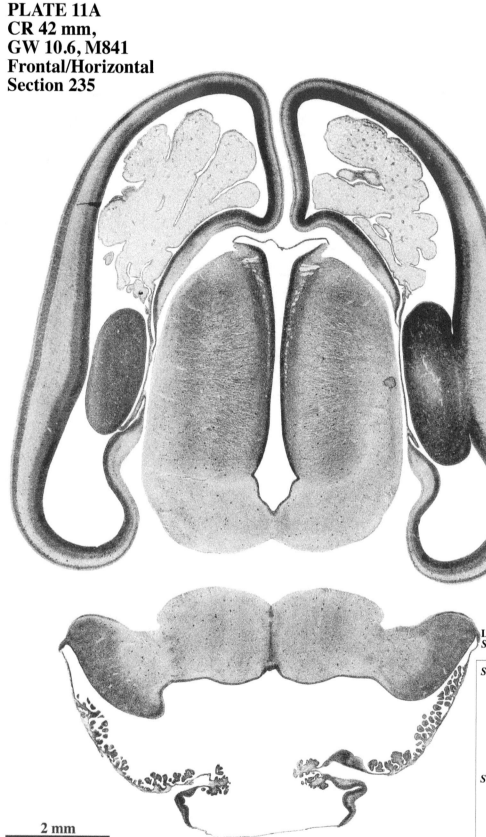

2 mm

LAYERS OF THE CORTICAL
STRATIFIED TRANSITIONAL
FIELD (STF)

STF1 Superficial fibrous layer, with an early developmental stage *(t1)* when many cells are migrating through it, followed by a late stage *(t2)* with sparse cells. Endures as the subcortical white matter.

STF4 Complex middle layer where sojourning and migrating cortical neurons grow corticofugal axons and intermingle with corticopetal axons.

STF5 Deep cellular layer that is prominent during the first trimester, the first sojourn zone to appear outside the germinal matrix.

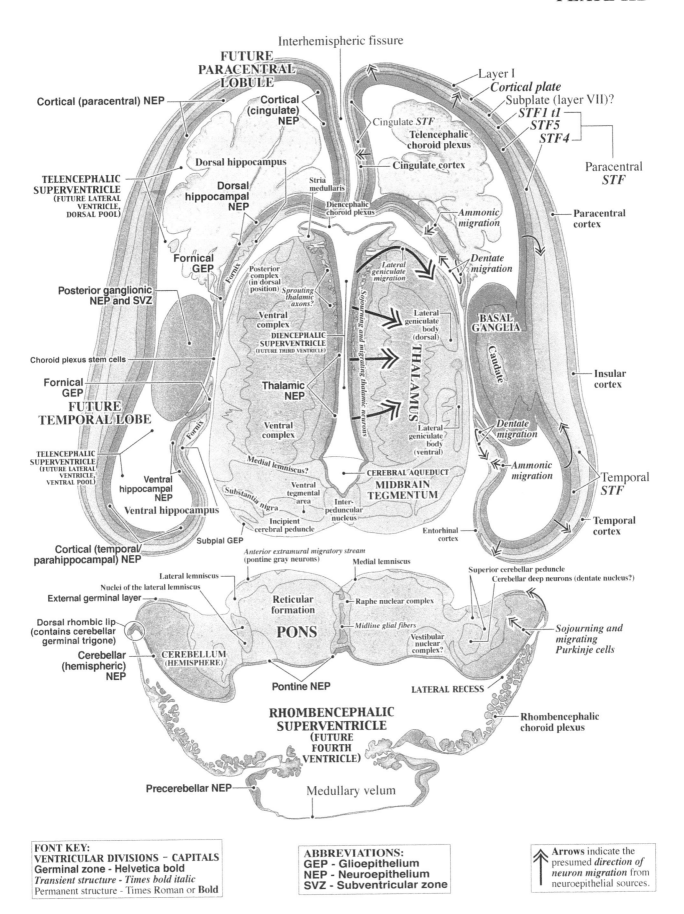

Interhemispheric fissure

FUTURE
PARACENTRAL
LOBULE

Cortical (paracentral) NEP

Cortical
(cingulate)
NEP

Cingulate *STF*

Layer I
Cortical plate
Subplate (layer VII)?
STF1 t1
STF5
STF4

Telencephalic
choroid plexus

Paracentral
STF

Dorsal hippocampus

Cingulate cortex

Dorsal
hippocampal
NEP

Stria
medullaris

Diencephalic
choroid plexus

*Ammonic
migration*

Paracentral
cortex

TELENCEPHALIC
SUPERVENTRICLE
(FUTURE LATERAL
VENTRICLE,
DORSAL POOL)

Fornical
GEP

Fornix

Posterior
complex
(in dorsal
position)

*Sprouting
thalamic
axons?*

*Lateral
geniculate
migration*

*Dentate
migration*

Posterior ganglionic
NEP and SVZ

Ventral
complex

Sojourning and migrating thalamic neurons

Lateral
geniculate
body
(dorsal)

BASAL
GANGLIA

Caudate

Choroid plexus stem cells

DIENCEPHALIC
SUPERVENTRICLE
(FUTURE THIRD VENTRICLE)

THALAMUS

Insular
cortex

Fornical
GEP

Thalamic
NEP

FUTURE
TEMPORAL LOBE

*Dentate
migration*

TELENCEPHALIC
SUPERVENTRICLE
(FUTURE LATERAL
VENTRICLE,
VENTRAL POOL)

Fornix

Ventral
complex

Lateral
geniculate
body
(ventral)

Ventral
hippocampal
NEP

*Ammonic
migration*

Temporal
STF

Ventral hippocampus

Medial lemniscus?

CEREBRAL AQUEDUCT
MIDBRAIN
TEGMENTUM

Substantia nigra

Ventral
tegmental
area

Inter-
peduncular
nucleus

Cortical (temporal/
parahippocampal) NEP

Subpial GEP

Incipient
cerebral peduncle

Entorhinal
cortex

Temporal
cortex

Anterior extramural migratory stream
(pontine gray neurons)

Medial lemniscus

Superior cerebellar peduncle
Cerebellar deep neurons (dentate nucleus?)

Lateral lemniscus

Nuclei of the lateral lemniscus

Reticular
formation

Raphe nuclear complex

External germinal layer

PONS

Midline glial fibers

Dorsal rhombic lip
(contains cerebellar
germinal trigone)

Vestibular
nuclear
complex?

*Sojourning and
migrating
Purkinje cells*

Cerebellar
(hemispheric)
NEP

CEREBELLUM
(HEMISPHERE)

Pontine NEP

LATERAL RECESS

Rhombencephalic
choroid plexus

RHOMBENCEPHALIC
SUPERVENTRICLE
(FUTURE
FOURTH
VENTRICLE)

Precerebellar NEP

Medullary velum

FONT KEY:
VENTRICULAR DIVISIONS – CAPITALS
Germinal zone - Helvetica bold
Transient structure - Times bold italic
Permanent structure - Times Roman or **Bold**

ABBREVIATIONS:
GEP - Glioepithelium
NEP - Neuroepithelium
SVZ - Subventricular zone

Arrows indicate the
presumed *direction of
neuron migration* from
neuroepithelial sources.

PLATE 12A
CR 42 mm,
GW 10.6, M841
Frontal/Horizontal
Section 220

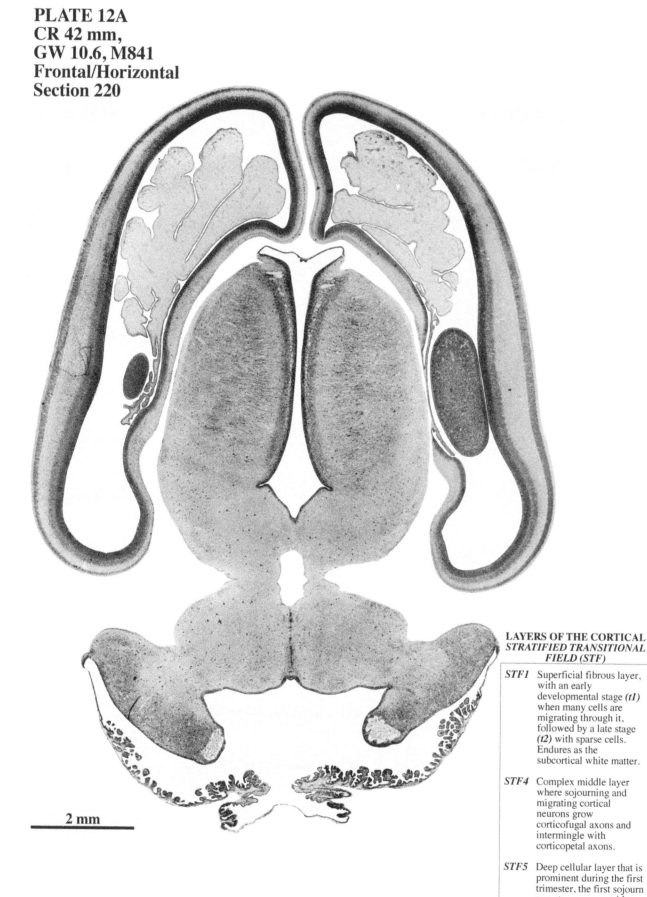

2 mm

LAYERS OF THE CORTICAL
STRATIFIED TRANSITIONAL
FIELD (STF)

STF1 Superficial fibrous layer, with an early developmental stage *(t1)* when many cells are migrating through it, followed by a late stage *(t2)* with sparse cells. Endures as the subcortical white matter.

STF4 Complex middle layer where sojourning and migrating cortical neurons grow corticofugal axons and intermingle with corticopetal axons.

STF5 Deep cellular layer that is prominent during the first trimester, the first sojourn zone to appear outside the germinal matrix.

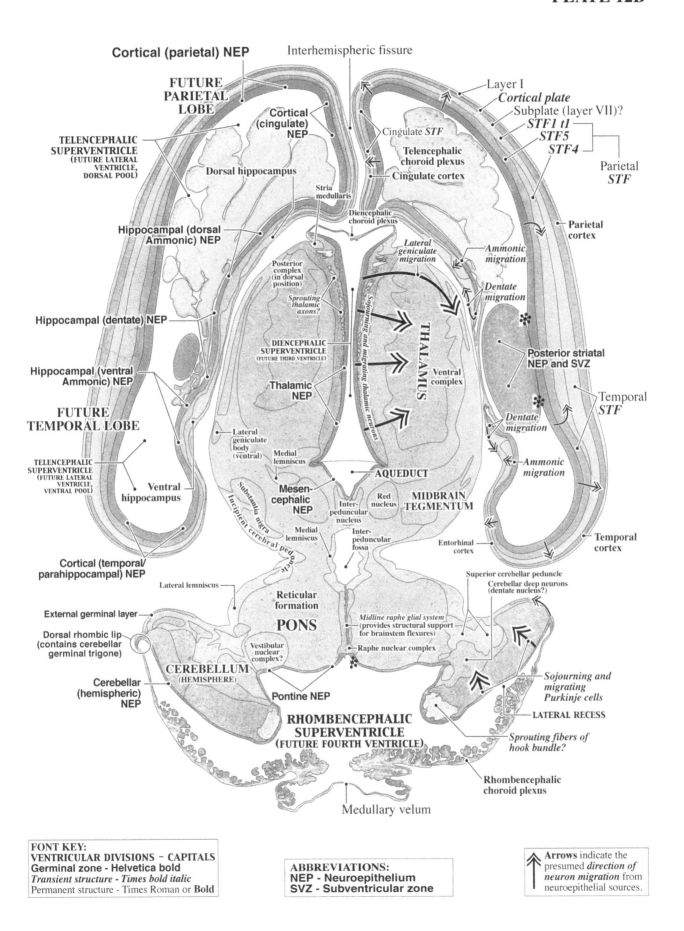

Cortical (parietal) NEP

Interhemispheric fissure

FUTURE PARIETAL LOBE

Cortical (cingulate) NEP

Layer I
Cortical plate
Subplate (layer VII)?
STF1 t1
STF5
STF4

Parietal *STF*

Cingulate *STF*

TELENCEPHALIC SUPERVENTRICLE
(FUTURE LATERAL VENTRICLE, DORSAL POOL)

Dorsal hippocampus

Telencephalic choroid plexus

Stria medullaris

Cingulate cortex

Diencephalic choroid plexus

Hippocampal (dorsal Ammonic) NEP

Posterior complex (in dorsal position)

Lateral geniculate migration

Ammonic migration

Parietal cortex

Sprouting thalamic axons?

Dentate migration

Hippocampal (dentate) NEP

DIENCEPHALIC SUPERVENTRICLE
(FUTURE THIRD VENTRICLE)

Sojourning and migrating thalamic neurons

THALAMUS

Posterior striatal NEP and SVZ

Hippocampal (ventral Ammonic) NEP

Thalamic NEP

Ventral complex

Temporal *STF*

FUTURE TEMPORAL LOBE

Lateral geniculate body (ventral)

Medial lemniscus

Dentate migration

TELENCEPHALIC SUPERVENTRICLE
(FUTURE LATERAL VENTRICLE, VENTRAL POOL)

Ventral hippocampus

Ammonic migration

AQUEDUCT

Substantia nigra

Incipient cerebral peduncle

Mesencephalic NEP

Inter-peduncular nucleus

Red nucleus

MIDBRAIN TEGMENTUM

Cortical (temporal/parahippocampal) NEP

Medial lemniscus

Inter-peduncular fossa

Entorhinal cortex

Temporal cortex

Lateral lemniscus

Reticular formation

PONS

Superior cerebellar peduncle

Cerebellar deep neurons (dentate nucleus?)

External germinal layer

Dorsal rhombic lip (contains cerebellar germinal trigone)

Vestibular nuclear complex?

Midline raphe glial system
(provides structural support for brainstem flexures)

Raphe nuclear complex

Sojourning and migrating Purkinje cells

CEREBELLUM
(HEMISPHERE)

LATERAL RECESS

Cerebellar (hemispheric) NEP

Pontine NEP

RHOMBENCEPHALIC SUPERVENTRICLE
(FUTURE FOURTH VENTRICLE)

Sprouting fibers of hook bundle?

Rhombencephalic choroid plexus

Medullary velum

FONT KEY:
VENTRICULAR DIVISIONS – CAPITALS
Germinal zone - Helvetica bold
Transient structure - Times bold italic
Permanent structure - Times Roman or **Bold**

ABBREVIATIONS:
NEP - Neuroepithelium
SVZ - Subventricular zone

Arrows indicate the presumed *direction of neuron migration* from neuroepithelial sources.

PLATE 13A
CR 42 mm,
GW 10.6, M841
Frontal/Horizontal
Section 205

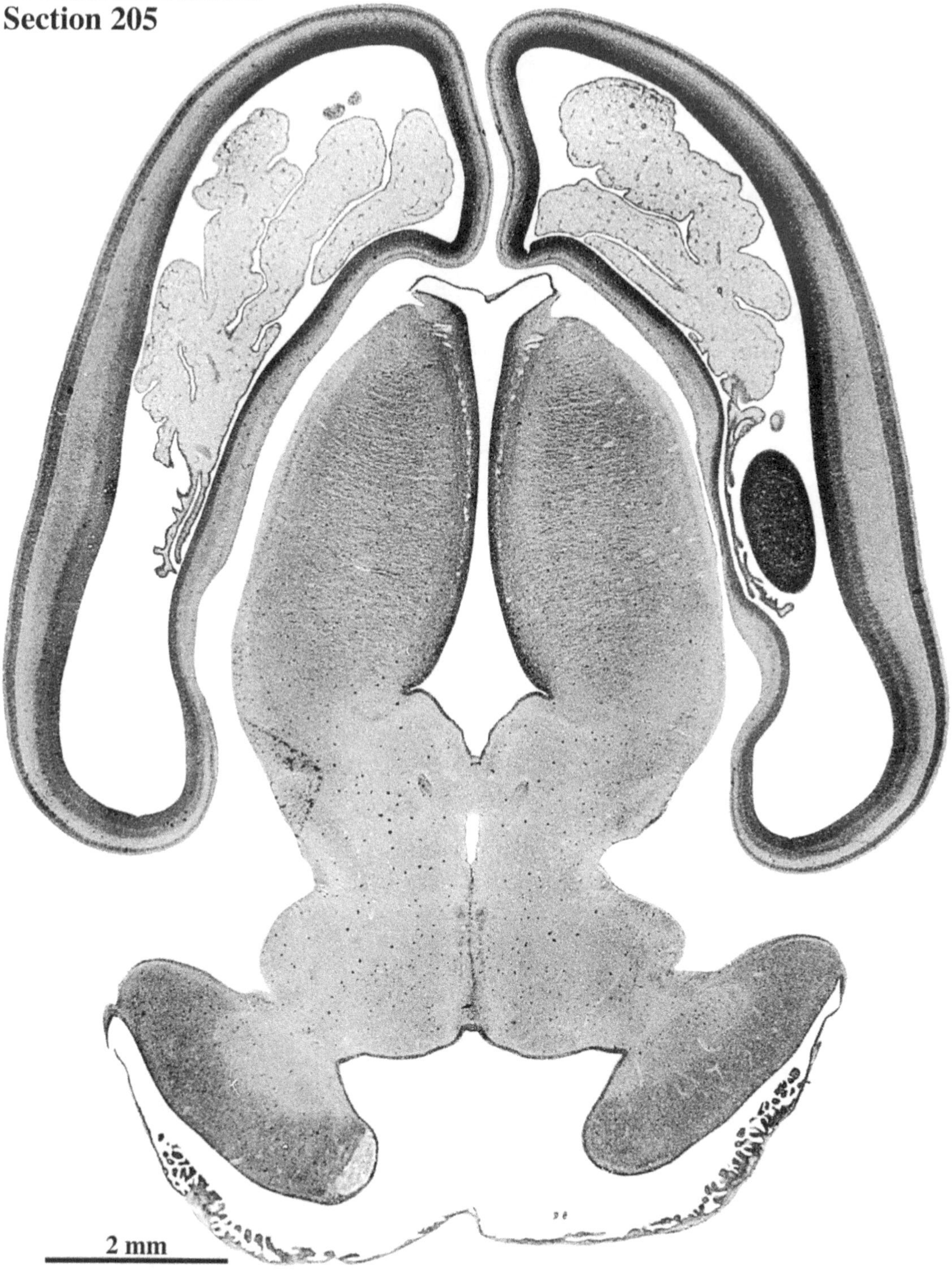

2 mm

LAYERS OF THE CORTICAL *STRATIFIED TRANSITIONAL FIELD (STF)*

STF1	Superficial fibrous layer, with an early developmental stage *(t1)* when many cells are migrating through it, followed by a late stage *(t2)* with sparse cells. Endures as the subcortical white matter.
STF4	Complex middle layer where sojourning and migrating cortical neurons grow corticofugal axons and intermingle with corticopetal axons.
STF5	Deep cellular layer that is prominent during the first trimester, the first sojourn zone to appear outside the germinal matrix.

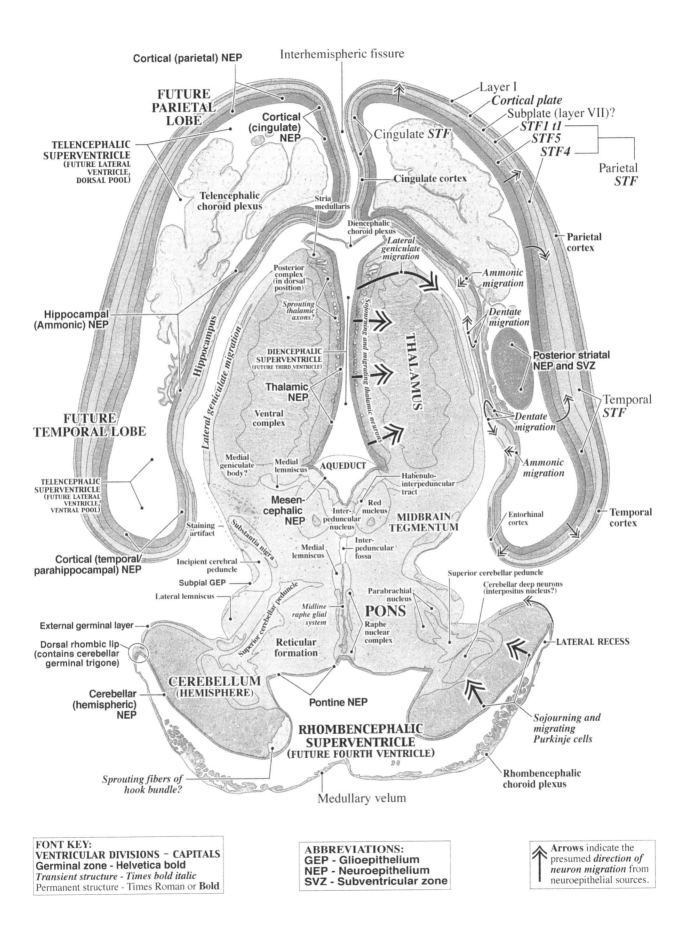

ABBREVIATIONS:
GEP - Glioepithelium
NEP - Neuroepithelium
SVZ - Subventricular zone

Arrows indicate the presumed *direction of neuron migration* from neuroepithelial sources.

PLATE 14A
CR 42 mm,
GW 10.6, M841
Frontal/Horizontal
Section 175

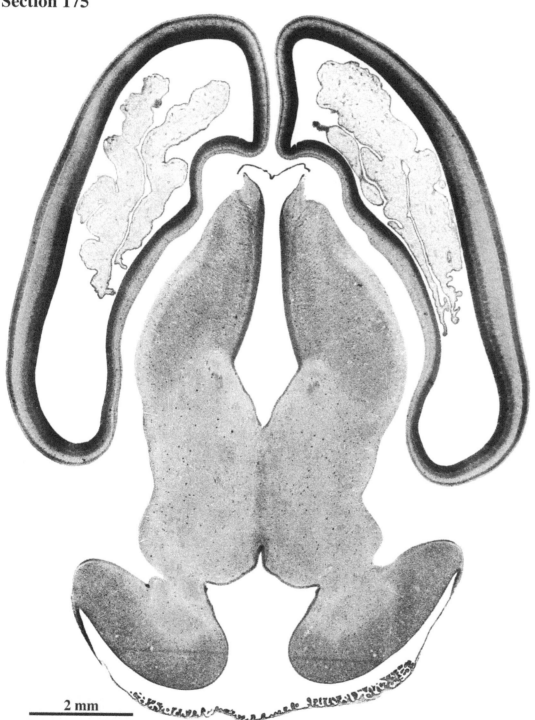

2 mm

LAYERS OF THE CORTICAL *STRATIFIED TRANSITIONAL FIELD (STF)*

STF1 Superficial fibrous layer, with an early developmental stage *(t1)* when many cells are migrating through it, followed by a late stage *(t2)* with sparse cells. Endures as the subcortical white matter.

STF4 Complex middle layer where sojourning and migrating cortical neurons grow corticofugal axons and intermingle with corticopetal axons.

STF5 Deep cellular layer that is prominent during the first trimester, the first sojourn zone to appear outside the germinal matrix.

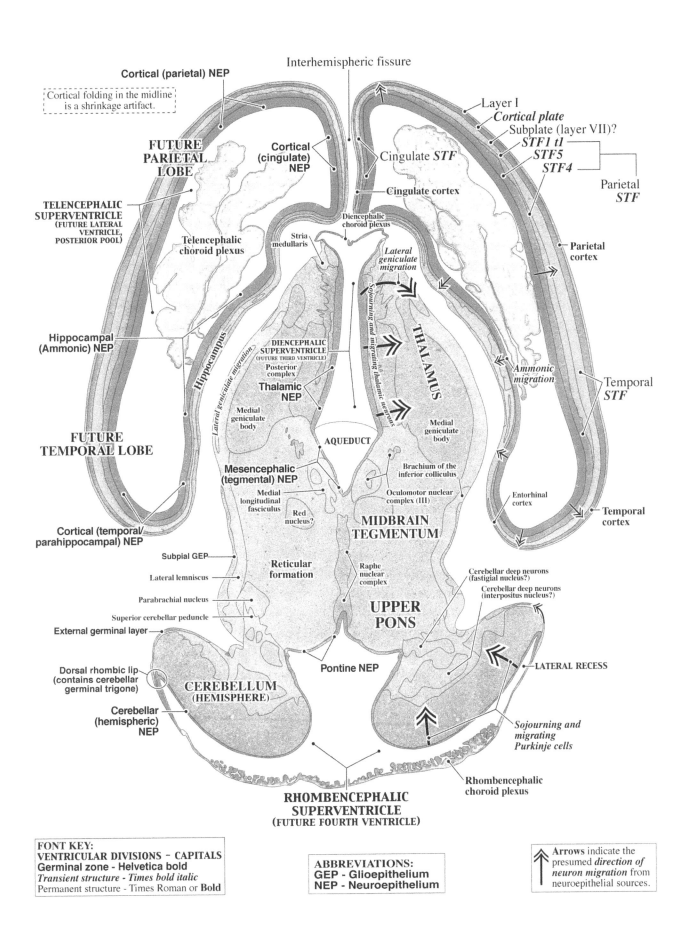

PLATE 15A
CR 42 mm,
GW 10.6, M841
Frontal/Horizontal
Section 155

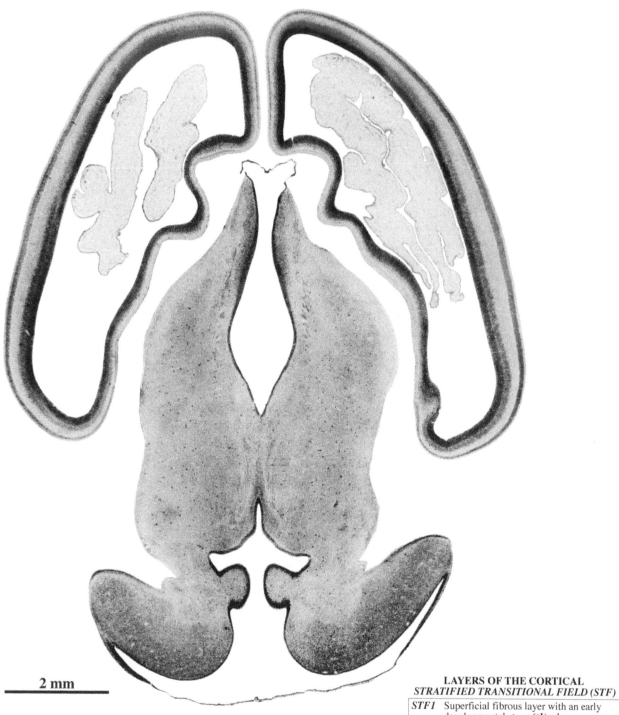

2 mm

LAYERS OF THE CORTICAL
STRATIFIED TRANSITIONAL FIELD (STF)

STF1	Superficial fibrous layer with an early developmental stage *(t1)* when many cells are migrating through it, followed by a late stage *(t2)* with sparse cells. Endures as the subcortical white matter.
STF5	Deep cellular layer that is prominent during the first trimester, the first sojourn zone to appear outside the germinal matrix.

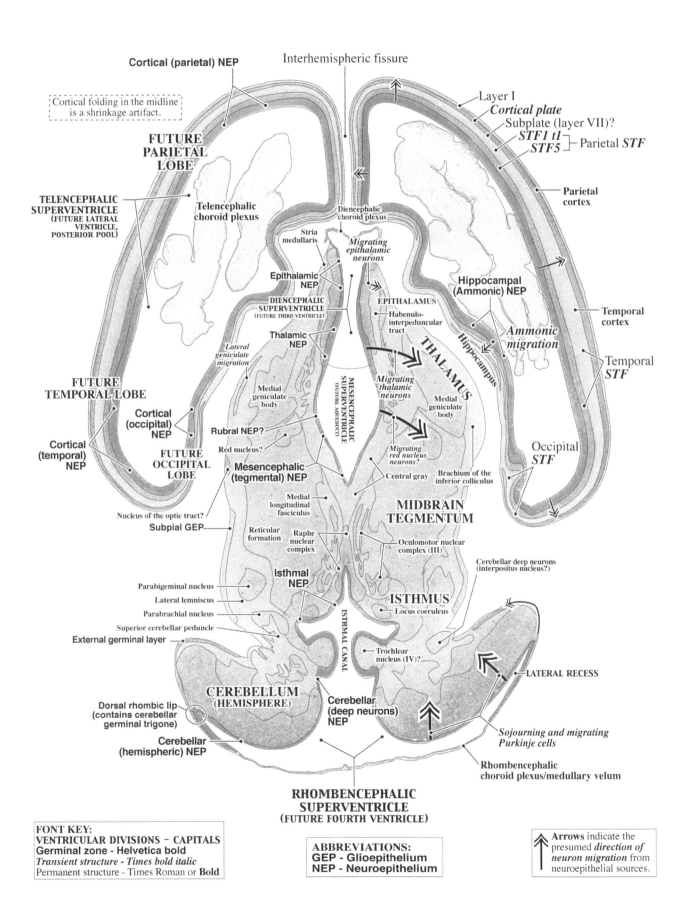

Interhemispheric fissure

Cortical (parietal) NEP

Cortical folding in the midline
is a shrinkage artifact.

**FUTURE
PARIETAL
LOBE**

Layer I
Cortical plate
Subplate (layer VII)?
STF1 t1
STF5] Parietal *STF*

Parietal
cortex

**TELENCEPHALIC
SUPERVENTRICLE**
(FUTURE LATERAL
VENTRICLE,
POSTERIOR POOL)

Telencephalic
choroid plexus

Diencephalic
choroid plexus

Stria
medullaris

*Migrating
epithalamic
neurons*

Epithalamic
NEP

Hippocampal
(Ammonic) NEP

**DIENCEPHALIC
SUPERVENTRICLE**
(FUTURE THIRD VENTRICLE)

EPITHALAMUS

Habenulo-
interpeduncular
tract

Temporal
cortex

Thalamic
NEP

*Ammonic
migration*

*Lateral
geniculate
migration*

THALAMUS

Hippocampus

Temporal
STF

Medial
geniculate
body

**MESENCEPHALIC
SUPERVENTRICLE**
(FUTURE AQUEDUCT)

*Migrating
thalamic
neurons*

Medial
geniculate
body

**FUTURE
TEMPORAL LOBE**

Cortical
(occipital)
NEP

Rubral NEP?

Red nucleus?

*Migrating
red nucleus
neurons?*

Occipital
STF

Cortical
(temporal)
NEP

**FUTURE
OCCIPITAL
LOBE**

**Mesencephalic
(tegmental) NEP**

Central gray

Brachium of the
inferior colliculus

Nucleus of the optic tract?

Medial
longitudinal
fasciculus

**MIDBRAIN
TEGMENTUM**

Subpial GEP

Reticular
formation

Raphe
nuclear
complex

Oculomotor nuclear
complex (III)

Cerebellar deep neurons
(interpositus nucleus?)

**Isthmal
NEP**

Parabigeminal nucleus

Lateral lemniscus

ISTHMUS

Locus coeruleus

Parabrachial nucleus

Superior cerebellar peduncle

External germinal layer

*Trochlear
nucleus (IV)?*

LATERAL RECESS

ISTHMAL CANAL

**CEREBELLUM
(HEMISPHERE)**

Cerebellar
(deep neurons)
NEP

Dorsal rhombic lip
(contains cerebellar
germinal trigone)

*Sojourning and migrating
Purkinje cells*

Cerebellar
(hemispheric) NEP

Rhombencephalic
choroid plexus/medullary velum

**RHOMBENCEPHALIC
SUPERVENTRICLE**
(FUTURE FOURTH VENTRICLE)

FONT KEY:
VENTRICULAR DIVISIONS – CAPITALS
Germinal zone - Helvetica bold
Transient structure - Times bold italic
Permanent structure - Times Roman or **Bold**

ABBREVIATIONS:
GEP - Glioepithelium
NEP - Neuroepithelium

Arrows indicate the
presumed *direction of
neuron migration* from
neuroepithelial sources.

PLATE 16A
CR 42 mm,
GW 10.6, M841
Frontal/Horizontal
Section 135

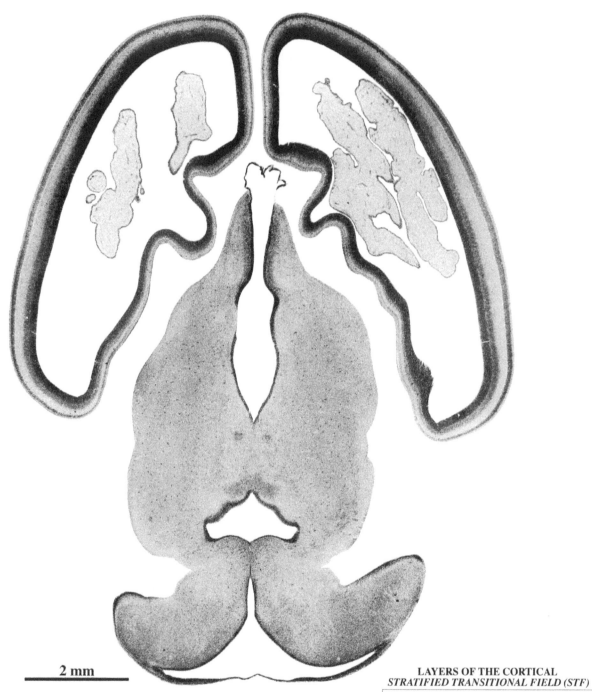

2 mm

LAYERS OF THE CORTICAL
STRATIFIED TRANSITIONAL FIELD (STF)

STF1	Superficial fibrous layer with an early developmental stage *(t1)* when many cells are migrating through it, followed by a late stage *(t2)* with sparse cells. Endures as the subcortical white matter.
STF5	Deep cellular layer that is prominent during the first trimester, the first sojourn zone to appear outside the germinal matrix.

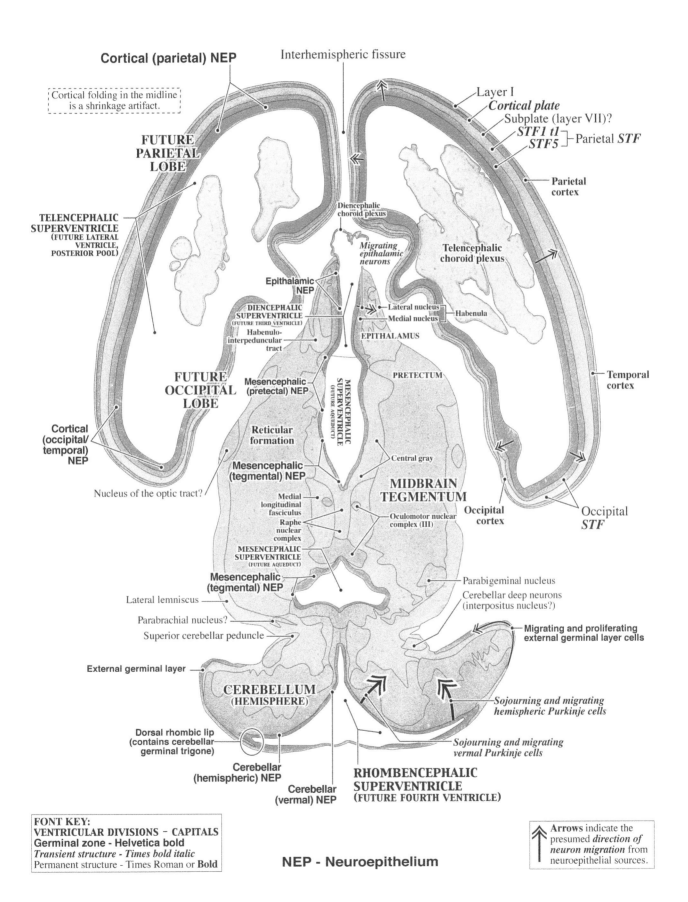

Cortical (parietal) NEP

Interhemispheric fissure

Cortical folding in the midline is a shrinkage artifact.

FUTURE PARIETAL LOBE

Layer I
Cortical plate
Subplate (layer VII)?
STF1 t1
STF5 — Parietal *STF*

Parietal cortex

TELENCEPHALIC SUPERVENTRICLE (FUTURE LATERAL VENTRICLE, POSTERIOR POOL)

Diencephalic choroid plexus

Migrating epithalamic neurons

Telencephalic choroid plexus

Epithalamic NEP

DIENCEPHALIC SUPERVENTRICLE (FUTURE THIRD VENTRICLE)

Lateral nucleus
Medial nucleus
Habenula

Habenulo-interpeduncular tract

EPITHALAMUS

FUTURE OCCIPITAL LOBE

Mesencephalic (pretectal) NEP

MESENCEPHALIC SUPERVENTRICLE (FUTURE AQUEDUCT)

PRETECTUM

Temporal cortex

Cortical (occipital/temporal) NEP

Reticular formation

Central gray

Nucleus of the optic tract?

Mesencephalic (tegmental) NEP

MIDBRAIN TEGMENTUM

Medial longitudinal fasciculus
Raphe nuclear complex

Oculomotor nuclear complex (III)

Occipital cortex

Occipital *STF*

MESENCEPHALIC SUPERVENTRICLE (FUTURE AQUEDUCT)

Mesencephalic (tegmental) NEP

Lateral lemniscus

Parabrachial nucleus?

Superior cerebellar peduncle

Parabigeminal nucleus
Cerebellar deep neurons (interpositus nucleus?)

Migrating and proliferating external germinal layer cells

External germinal layer

CEREBELLUM (HEMISPHERE)

Sojourning and migrating hemispheric Purkinje cells

Dorsal rhombic lip (contains cerebellar germinal trigone)

Cerebellar (hemispheric) NEP

Cerebellar (vermal) NEP

Sojourning and migrating vermal Purkinje cells

RHOMBENCEPHALIC SUPERVENTRICLE (FUTURE FOURTH VENTRICLE)

FONT KEY:
VENTRICULAR DIVISIONS – CAPITALS
Germinal zone - Helvetica bold
Transient structure - Times bold italic
Permanent structure - Times Roman or **Bold**

NEP - Neuroepithelium

Arrows indicate the presumed *direction of neuron migration* from neuroepithelial sources.

PLATE 17A
CR 42 mm,
GW 10.6, M841
Frontal/Horizontal
Section 124

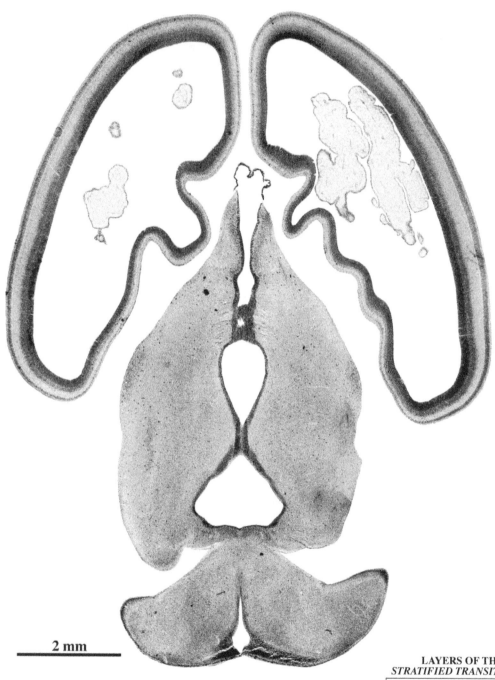

2 mm

LAYERS OF THE CORTICAL
STRATIFIED TRANSITIONAL FIELD (STF)

STF1	Superficial fibrous layer with an early developmental stage *(t1)* when many cells are migrating through it, followed by a late stage *(t2)* with sparse cells. Endures as the subcortical white matter.
STF5	Deep cellular layer that is prominent during the first trimester, the first sojourn zone to appear outside the germinal matrix.

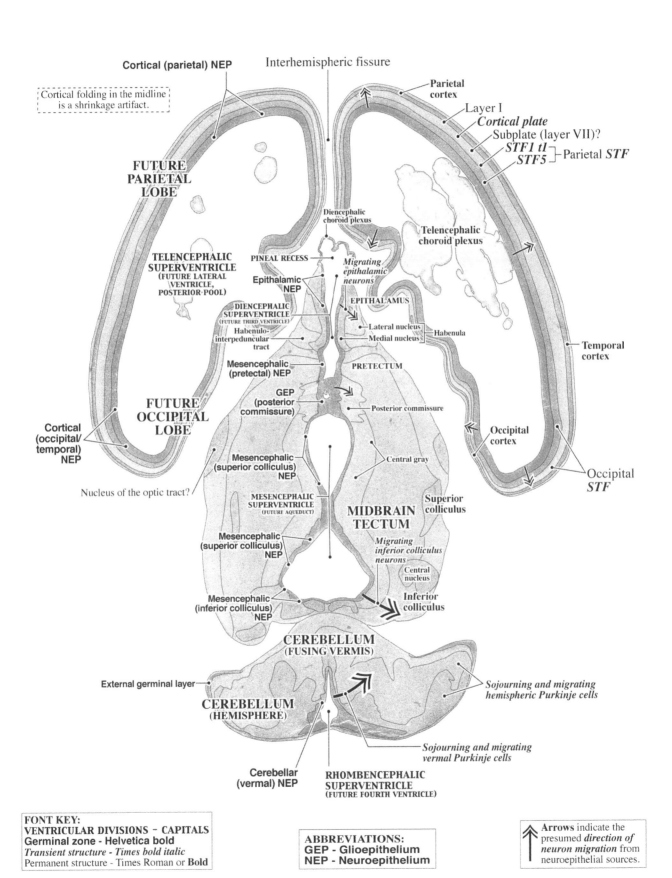

Cortical (parietal) NEP

Interhemispheric fissure

Parietal cortex

Cortical folding in the midline is a shrinkage artifact.

Layer I
Cortical plate
Subplate (layer VII)?
STF1 t1
STF5 Parietal *STF*

FUTURE PARIETAL LOBE

Diencephalic choroid plexus

Telencephalic choroid plexus

TELENCEPHALIC SUPERVENTRICLE
(FUTURE LATERAL VENTRICLE, POSTERIOR POOL)

PINEAL RECESS

Migrating epithalamic neurons

Epithalamic **NEP**

DIENCEPHALIC SUPERVENTRICLE
(FUTURE THIRD VENTRICLE)

EPITHALAMUS

Habenulo-interpeduncular tract

Lateral nucleus

Medial nucleus

Habenula

Mesencephalic (pretectal) **NEP**

PRETECTUM

Temporal cortex

FUTURE OCCIPITAL LOBE

GEP (posterior commissure)

Posterior commissure

Cortical (occipital/ temporal) **NEP**

Central gray

Occipital cortex

Mesencephalic (superior colliculus) **NEP**

Nucleus of the optic tract?

Occipital *STF*

MESENCEPHALIC SUPERVENTRICLE
(FUTURE AQUEDUCT)

Superior colliculus

MIDBRAIN TECTUM

Mesencephalic (superior colliculus) **NEP**

Migrating inferior colliculus neurons

Central nucleus

Mesencephalic (inferior colliculus) **NEP**

Inferior colliculus

CEREBELLUM (FUSING VERMIS)

External germinal layer

Sojourning and migrating hemispheric Purkinje cells

CEREBELLUM (HEMISPHERE)

Sojourning and migrating vermal Purkinje cells

Cerebellar (vermal) **NEP**

RHOMBENCEPHALIC SUPERVENTRICLE
(FUTURE FOURTH VENTRICLE)

FONT KEY:
VENTRICULAR DIVISIONS – CAPITALS
Germinal zone - Helvetica bold
Transient structure - Times bold italic
Permanent structure - Times Roman or **Bold**

ABBREVIATIONS:
GEP - Glioepithelium
NEP - Neuroepithelium

Arrows indicate the presumed *direction of neuron migration* from neuroepithelial sources.

PLATE 18A
CR 42 mm,
GW 10.6, M841
Frontal/Horizontal
Section 112

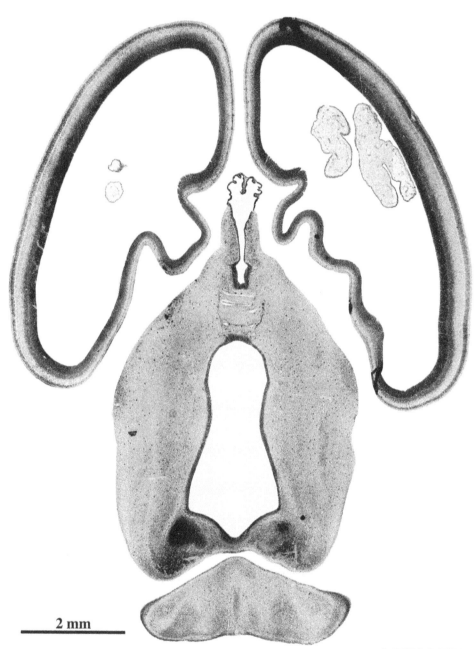

2 mm

LAYERS OF THE CORTICAL
STRATIFIED TRANSITIONAL FIELD (STF)

STF1 Superficial fibrous layer with an early
developmental stage *(t1)* when many
cells are migrating through it, followed
by a late stage *(t2)* with sparse cells.
Endures as the subcortical white matter.

STF5 Deep cellular layer that is prominent
during the first trimester, the first sojourn
zone to appear outside the germinal
matrix.

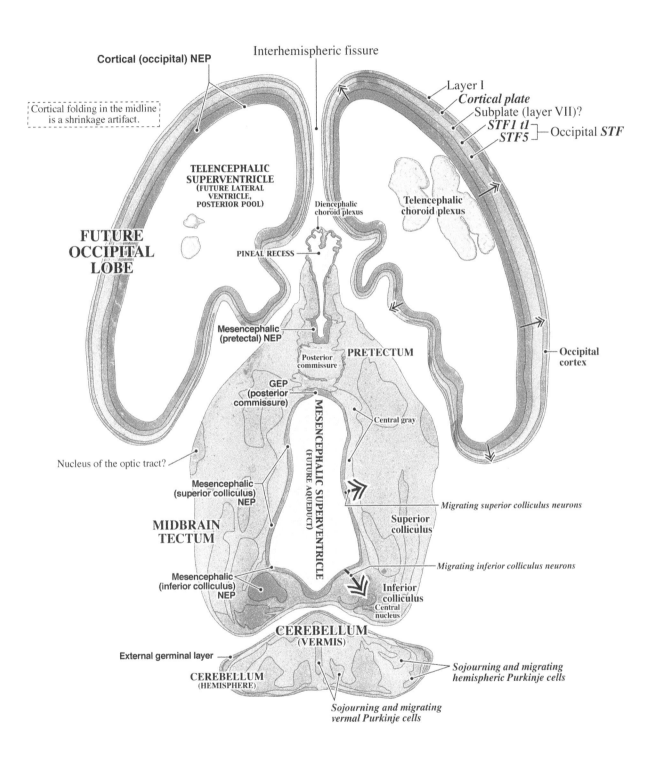

Cortical (occipital) NEP

Interhemispheric fissure

Layer I
Cortical plate
Subplate (layer VII)?
STF1 t1
STF5 — Occipital *STF*

Cortical folding in the midline
is a shrinkage artifact.

**TELENCEPHALIC
SUPERVENTRICLE**
(FUTURE LATERAL
VENTRICLE,
POSTERIOR POOL)

Diencephalic
choroid plexus

**Telencephalic
choroid plexus**

**FUTURE
OCCIPITAL
LOBE**

PINEAL RECESS

Mesencephalic
(pretectal) NEP

PRETECTUM

Posterior
commissure

GEP
(posterior
commissure)

Central gray

**Occipital
cortex**

Nucleus of the optic tract?

Mesencephalic
(superior colliculus)
NEP

MESENCEPHALIC SUPERVENTRICLE
(FUTURE AQUEDUCT)

Migrating superior colliculus neurons

**Superior
colliculus**

**MIDBRAIN
TECTUM**

Mesencephalic
(inferior colliculus)
NEP

Migrating inferior colliculus neurons

**Inferior
colliculus**
Central
nucleus

CEREBELLUM
(VERMIS)

External germinal layer

*Sojourning and migrating
hemispheric Purkinje cells*

CEREBELLUM
(HEMISPHERE)

*Sojourning and migrating
vermal Purkinje cells*

FONT KEY:
VENTRICULAR DIVISIONS – CAPITALS
Germinal zone - Helvetica bold
Transient structure - Times bold italic
Permanent structure - Times Roman or **Bold**

ABBREVIATIONS:
GEP - Glioepithelium
NEP - Neuroepithelium

Arrows indicate the
presumed *direction of
neuron migration* from
neuroepithelial sources.

PLATE 19A
CR 42 mm,
GW 10.6, M841
Frontal/Horizontal
Section 96

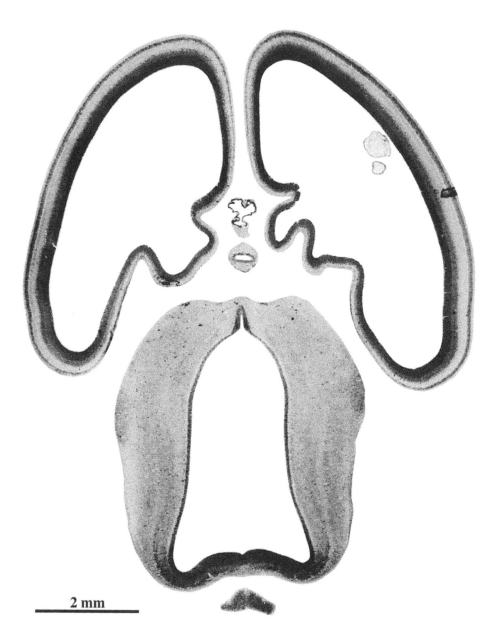

2 mm

LAYERS OF THE CORTICAL
STRATIFIED TRANSITIONAL FIELD (STF)

STF1	Superficial fibrous layer with an early developmental stage *(t1)* when many cells are migrating through it, followed by a late stage *(t2)* with sparse cells. Endures as the subcortical white matter.
STF5	Deep cellular layer that is prominent during the first trimester, the first sojourn zone to appear outside the germinal matrix.

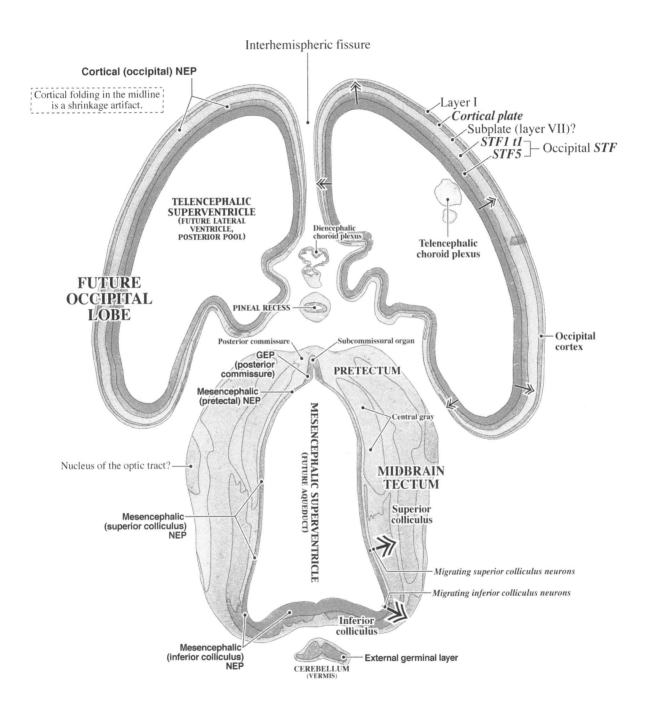

Interhemispheric fissure

Cortical (occipital) NEP

Cortical folding in the midline
is a shrinkage artifact.

Layer I
Cortical plate
Subplate (layer VII)?
STF1 t1
STF5 — Occipital *STF*

**TELENCEPHALIC
SUPERVENTRICLE**
(FUTURE LATERAL
VENTRICLE,
POSTERIOR POOL)

Diencephalic
choroid plexus

Telencephalic
choroid plexus

**FUTURE
OCCIPITAL
LOBE**

PINEAL RECESS

Occipital
cortex

Posterior commissure   Subcommissural organ

GEP
(posterior
commissure)

PRETECTUM

**Mesencephalic
(pretectal) NEP**

**MESENCEPHALIC
SUPERVENTRICLE**
(FUTURE AQUEDUCT)

Central gray

Nucleus of the optic tract?

**MIDBRAIN
TECTUM**

Superior
colliculus

**Mesencephalic
(superior colliculus)
NEP**

Migrating superior colliculus neurons

Migrating inferior colliculus neurons

Inferior
colliculus

**Mesencephalic
(inferior colliculus)
NEP**

External germinal layer

CEREBELLUM
(VERMIS)

FONT KEY:
VENTRICULAR DIVISIONS – CAPITALS
Germinal zone - Helvetica bold
Transient structure - Times bold italic
Permanent structure - Times Roman or **Bold**

ABBREVIATIONS:
GEP - Glioepithelium
NEP - Neuroepithelium

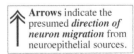

Arrows indicate the
presumed *direction of
neuron migration* from
neuroepithelial sources.

PLATE 20A
CR 42 mm,
GW 10.6, M841
Frontal/Horizontal
Section 39

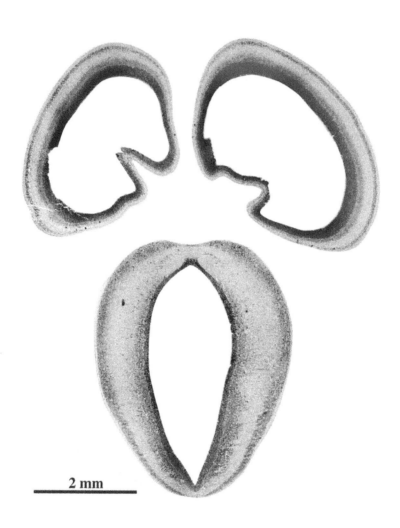

2 mm

LAYERS OF THE CORTICAL
STRATIFIED TRANSITIONAL FIELD (STF)

STF1 Superficial fibrous layer with an early
developmental stage *(t1)* when many
cells are migrating through it, followed
by a late stage *(t2)* with sparse cells.
Endures as the subcortical white matter.

STF5 Deep cellular layer that is prominent
during the first trimester, the first sojourn
zone to appear outside the germinal
matrix.

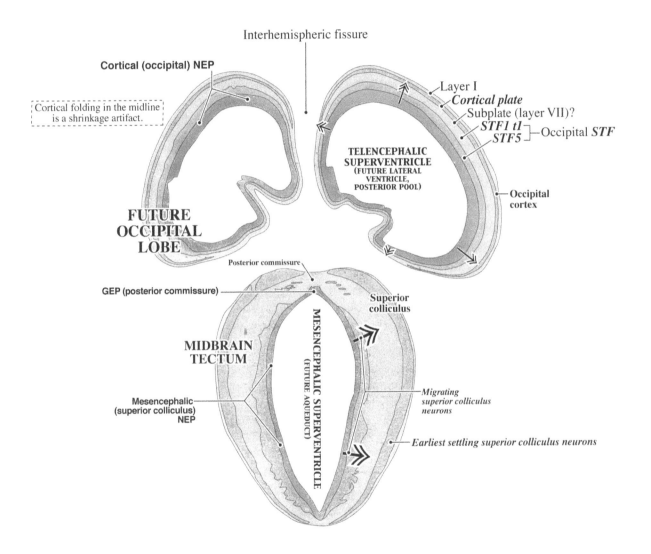

Interhemispheric fissure

Cortical (occipital) NEP

Cortical folding in the midline is a shrinkage artifact.

Layer I
Cortical plate
Subplate (layer VII)?
STF1 t1
STF5 — Occipital *STF*

TELENCEPHALIC SUPERVENTRICLE
(FUTURE LATERAL VENTRICLE, POSTERIOR POOL)

Occipital cortex

FUTURE OCCIPITAL LOBE

Posterior commissure

GEP (posterior commissure)

Superior colliculus

MIDBRAIN TECTUM

MESENCEPHALIC SUPERVENTRICLE
(FUTURE AQUEDUCT)

Mesencephalic (superior colliculus) NEP

Migrating superior colliculus neurons

Earliest settling superior colliculus neurons

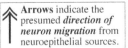

PART III: C886
CR 42 mm (GW 10.6)
Horizontal

This is specimen number 886 in the Carnegie collection, designated here as C886. A normal fetus with a crown rump length (CR) of 43 mm was collected in 1914 after a hysterectomy due to pelvic inflammation. The fetus is estimated to be at gestational week (GW) 10.6. The entire fetus was fixed in Bouin's, embedded in celloidin, and 100 μm sections were cut in a plane midway between the horizontal and frontal planes (**Fig. 10**). Because many of the sections do not contain the cerebral cortex and the brainstem is cut in a crosswise direction, it more closely resembles a horizontally-sectioned brain. All sections were stained with hematoxylin and eosin. Since there is no photograph of C886's brain before it was embedded and cut, a specimen from Hochstetter (1919) is used to show the external features of the brain at GW10.6 (**Fig. 10**). Larger sections containing the cerebral hemispheres, are shown at low magnification in **Plates 21A–B** to **30A–B**. The core parts of the sections in **Plates 23C–D** to **30C–D** are also shown at slightly higher magnification. **Plates 31A–B** to **38A–B** are shown only at high magnification. To more efficiently use page space, all plates are in landscape orientation (anterioventral: left side of photograph; posteriodorsal: right side of photograph).

C886 is similar to C6658 in the level of brain maturation. The chief reason for including this specimen is to provide a different perspective for viewing brain structures at GW10.6. In the cerebral cortex, the neuroepithelium appears to be the sole germinal matrix, but no doubt there is a subventricular zone in more mature lateral areas. The anteroateral (thicker) to dorsomedial (thinner) maturation gradient in the cortical plate and layers of the stratified transitional field (STF) are prominent. STF1 and STF5 are in in all areas, and STF4 is only in lateral areas. In anterolateral parts of the cerebral cortex, streams of neurons and glia appear to enter STF4 and join the lateral migratory stream. The hippocampus contains ammonic and dentate migrations, but there is no evidence of a pyramidal layer in Ammon's horn or of a granular layer dentate gyrus. A massive neuroepithelium/subventricular zone overlies the amygdala, nucleus accumbens, and striatum (caudate and putamen) where neurons (and glia) are being generated.

The neuroepithelium is still active in the hypothalamus and thalamus, where the last neuronal populations in these brain regions are being generated. Neurons throughout the diencephalon, are migrating and settling. This specimen shows a prominent migration of subthalamic nucleus neurons from the posterior hypothalamic neuroepithelium. Except for the subthalamic nucleus, nuclear divisions are very indistinct throughout the diencephalon because migrating neurons blur nuclear borders.

The floor of the aqueduct, and fourth ventricle are lined by thin neuroepithelia that are being transitioned to an ependymal layer. The two exceptions are the midtrain tectum and the precerebellar neuroepithelium in the upper medulla. These are the only germinal zones still producing neurons in the entire brainstem. Many neurons have already settled throughout the pons and medulla, only a few are still migrating. That allows more nuclear definition in these brain regions. However, the anterior and posterior extramural migratory streams are dense subpial accumulations in the medulla and pons, so structures like the lateral reticular and external cuneate nuclei in the medulla are just beginning to be visible. The pontine nuclei at the base of the pons are still absent, but dense accumulations of pontine nuclear neurons are easily visible on their way to their final settling place.

The cerebellum is a thick, smooth plate overlying the pons and upper medulla, with a thin neuroepithelium at the ventricular surface. Many Purkinje cells are sojourning in a dense layer outside the neuroepithelium, making the neuroepithelium appear thick. Older Purkinje cells are migrating upward to settle in the primordial cortex beneath the forward-growing external germinal layer (egl) that lies at the pial surface. Some of the deep neurons are still superficial in the cerebellum, but many are migrating downward to intermingle with upwardly migrating Purkinje cells.

GW10.6 HORIZONTAL/FRONTAL SECTION PLANES

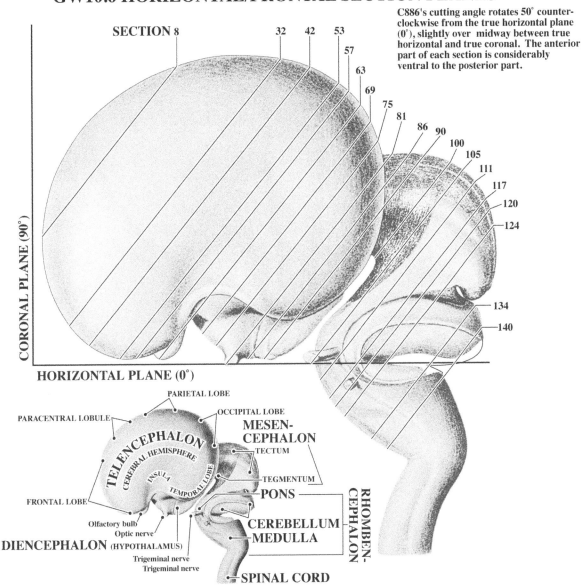

SECTION 8   32   42   53
57
63
69
75
81
86   90
100
105
111
117
120
124
134
140

C886's cutting angle rotates 50° counter-clockwise from the true horizontal plane (0°), slightly over midway between true horizontal and true coronal. The anterior part of each section is considerably ventral to the posterior part.

CORONAL PLANE (90°)

HORIZONTAL PLANE (0°)

PARIETAL LOBE
PARACENTRAL LOBULE
OCCIPITAL LOBE
MESEN-CEPHALON
TELENCEPHALON
CEREBRAL HEMISPHERE
TECTUM
INSULA
TEMPORAL LOBE
TEGMENTUM
PONS
FRONTAL LOBE
RHOMBEN-CEPHALON
Olfactory bulb
Optic nerve
CEREBELLUM
MEDULLA
DIENCEPHALON (HYPOTHALAMUS)
Trigeminal nerve
Trigeminal nerve
SPINAL CORD

Figure 10. The lateral view of the brain and upper cervical spinal cord from a specimen with a crown rump length of 38 mm (modified from Figure 43, Table VII, Hochstetter, 1919) serves to show the approximate locations and cutting angles of the illustrated sections of C886 in the following pages. The small inset identifies the major structural features. The line in the cerebellum and dorsal edges of the pons and medulla is the cut edge of the medullary velum.

PLATE 21A
CR 43 mm, GW10.6, C886, Horizontal/Frontal

Section 8

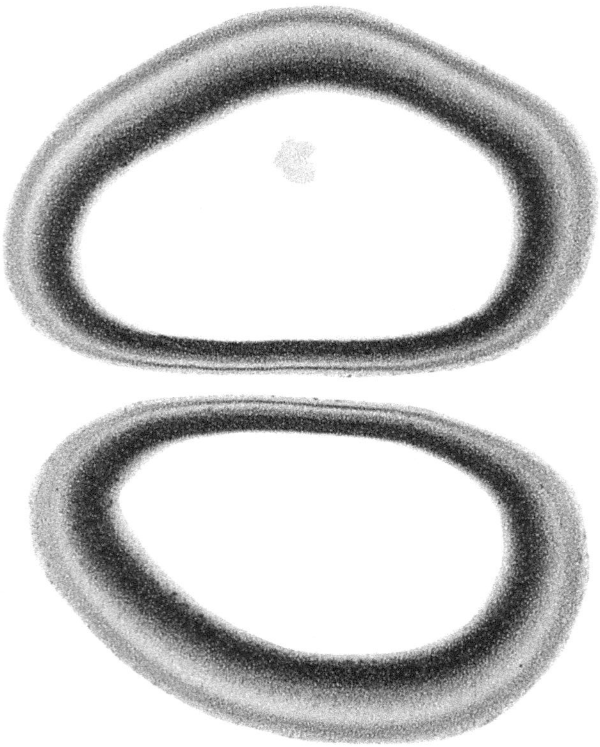

Section 33

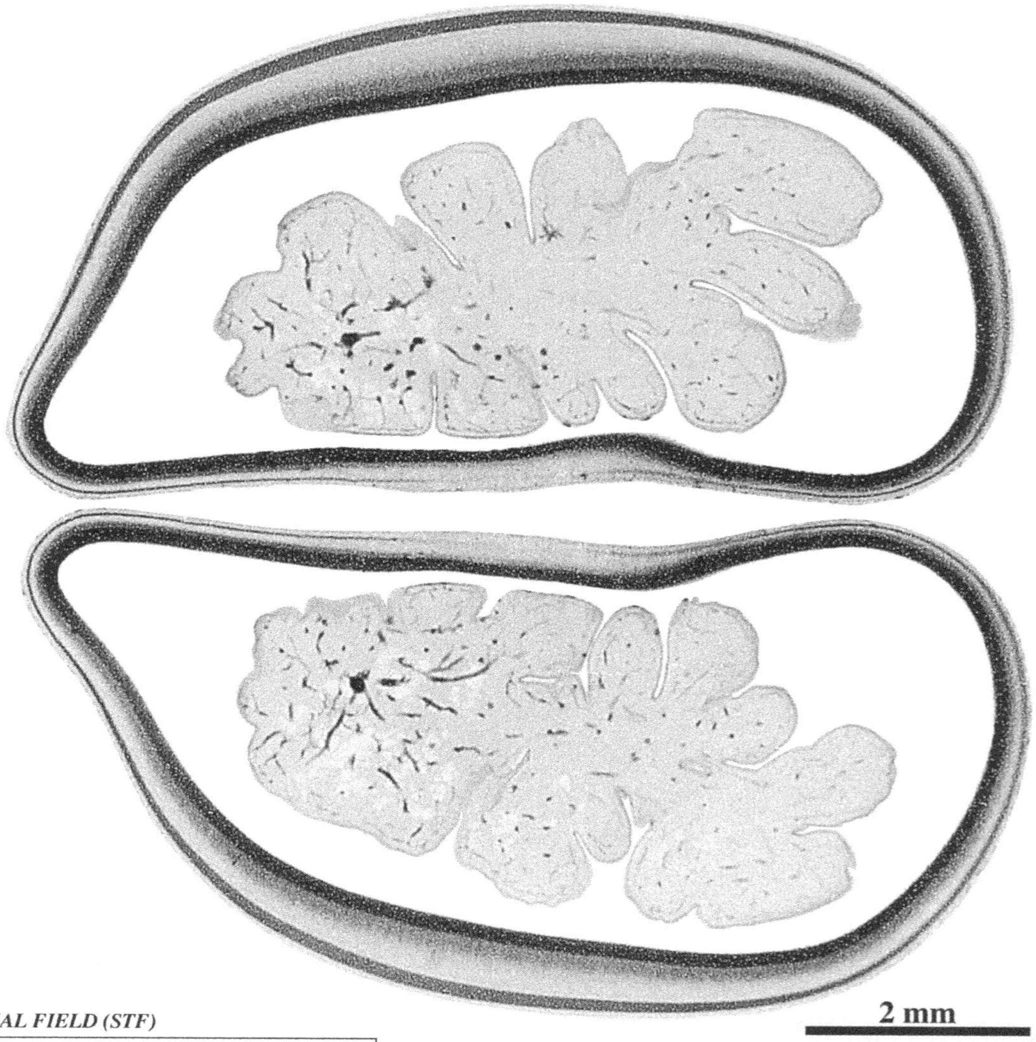

2 mm

LAYERS OF THE CORTICAL *STRATIFIED TRANSITIONAL FIELD (STF)*

STF1 Superficial fibrous layer with an early developmental stage (*t1*) when many cells are migrating through it, followed by a late stage (*t2*) with sparse cells. Endures as the subcortical white matter.

STF5 Deep cellular layer that is prominent during the first trimester, the first sojourn zone to appear outside the germinal matrix.

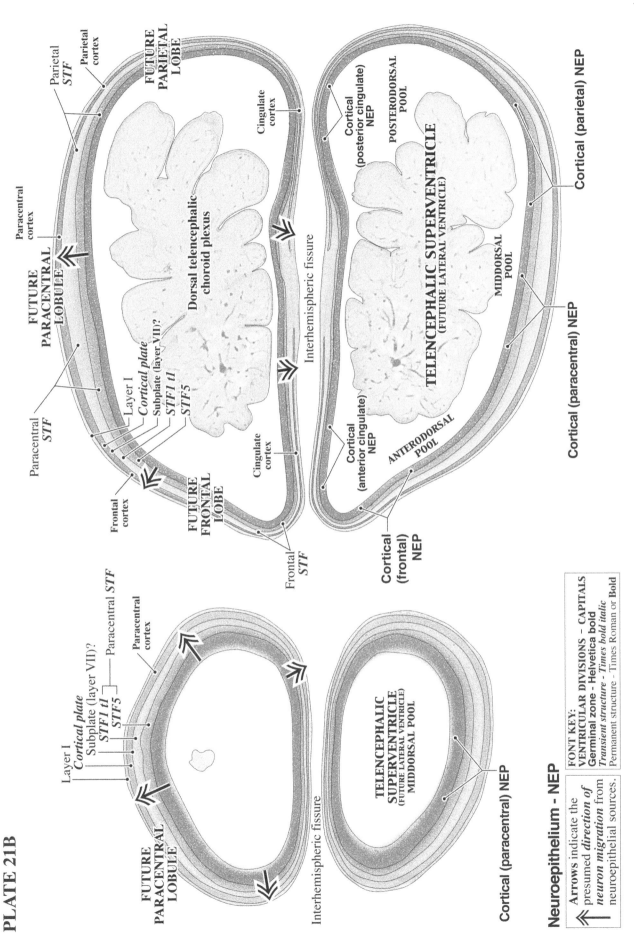

PLATE 21B

59

Upper figure labels:

Parietal *STF*

Parietal cortex

FUTURE PARIETAL LOBE

Paracentral cortex

FUTURE PARACENTRAL LOBULE

Paracentral *STF*

Layer I
Cortical plate
Subplate (layer VII)?
STF1 t1
STF5

Frontal cortex

FUTURE FRONTAL LOBE

Frontal *STF*

Dorsal telencephalic choroid plexus

Interhemispheric fissure

Cingulate cortex

Cingulate cortex

Cortical (posterior cingulate) NEP

POSTERODORSAL POOL

TELENCEPHALIC SUPERVENTRICLE
(FUTURE LATERAL VENTRICLE)

MIDDORSAL POOL

Cortical (anterior cingulate) NEP

ANTERODORSAL POOL

Cortical (frontal) NEP

Cortical (parietal) NEP

Cortical (paracentral) NEP

Cortical (paracentral) NEP

Lower figure labels:

Paracentral *STF*

Paracentral cortex

Layer I
Cortical plate
Subplate (layer VII)?
STF1 t1
STF5

FUTURE PARACENTRAL LOBULE

Interhemispheric fissure

TELENCEPHALIC SUPERVENTRICLE
(FUTURE LATERAL VENTRICLE)
MIDDORSAL POOL

Cortical (paracentral) NEP

Neuroepithelium - NEP

⇐ Arrows indicate the presumed *direction of neuron migration* from neuroepithelial sources.

FONT KEY:
VENTRICULAR DIVISIONS – CAPITALS
Germinal zone - **Helvetica bold**
Transient structure - *Times bold italic*
Permanent structure - Times Roman or **Bold**

PLATE 22A
CR 43 mm, GW10.6, C886
Horizontal/Frontal
Section 42

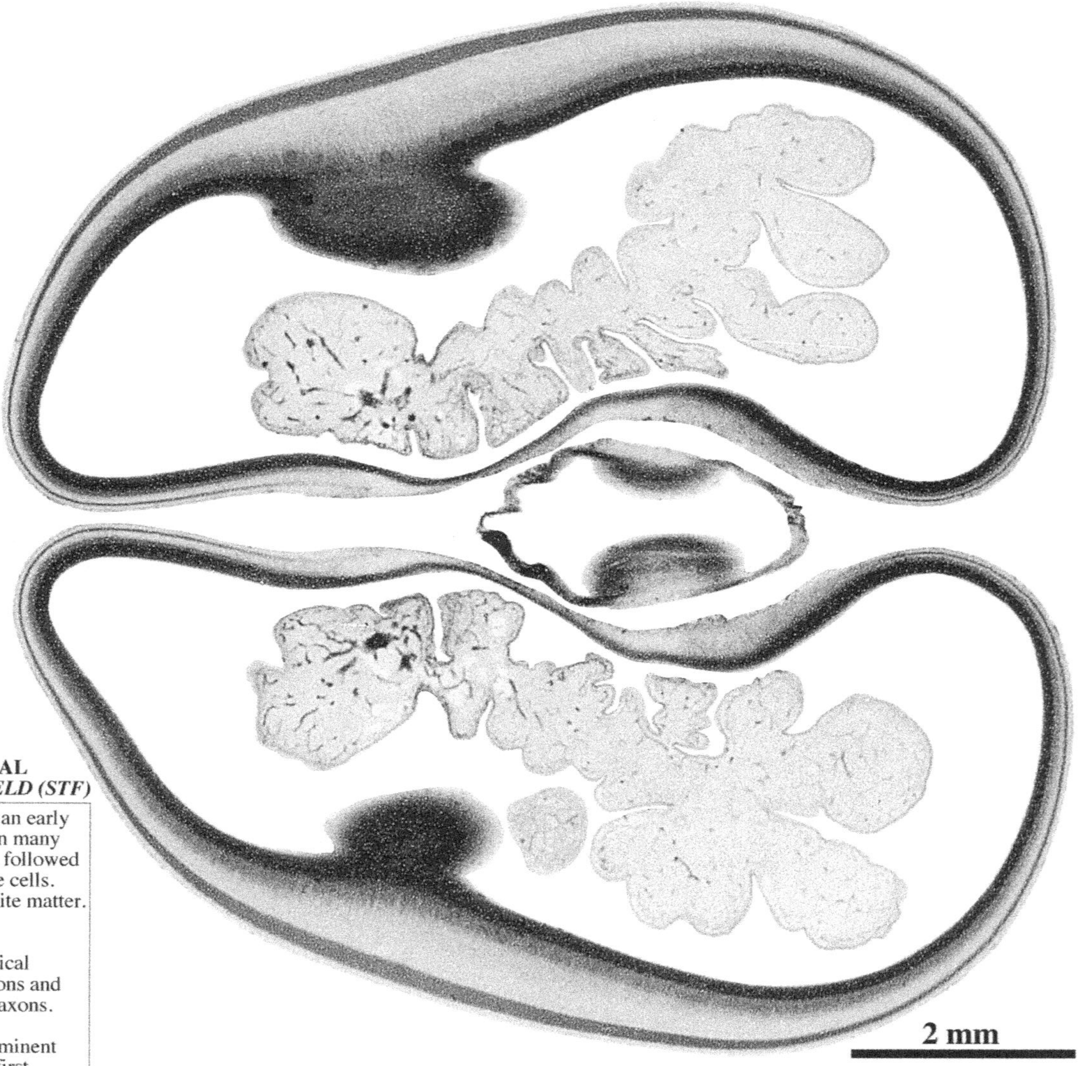

2 mm

LAYERS OF THE CORTICAL
STRATIFIED TRANSITIONAL FIELD (STF)

STF1   Superficial fibrous layer with an early
developmental stage (*t1*) when many
cells are migrating through it, followed
by a late stage (*t2*) with sparse cells.
Endures as the subcortical white matter.

STF4   Complex middle layer where
sojourning and migrating cortical
neurons grow corticofugal axons and
intermingle with corticopetal axons.

STF5   Deep cellular layer that is prominent
during the first trimester, the first
sojourn zone to appear outside the
germinal matrix.

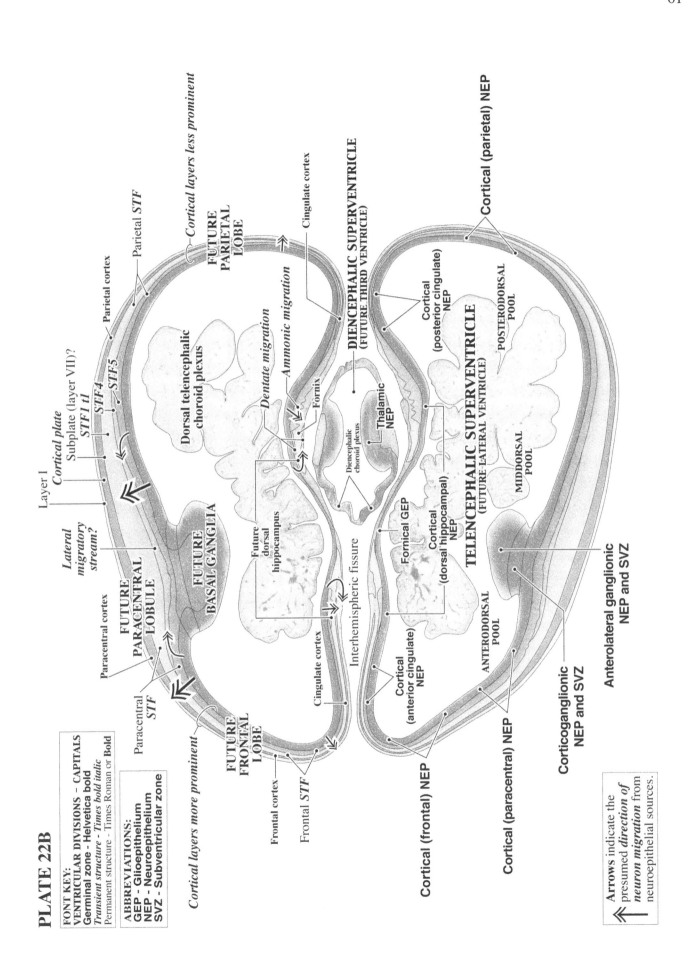

61

PLATE 22B

FONT KEY:
VENTRICULAR DIVISIONS – CAPITALS
Germinal zone - **Helvetica bold**
Transient structure - *Times bold italic*
Permanent structure - Times Roman or **Bold**

ABBREVIATIONS:
GEP - Glioepithelium
NEP - Neuroepithelium
SVZ - Subventricular zone

Layer 1

Cortical plate

Subplate (layer VII)?

STF1 t1

STF4

STF5

Parietal *STF*

Parietal cortex

Cortical layers less prominent

FUTURE PARIETAL LOBE

Cingulate cortex

DIENCEPHALIC SUPERVENTRICLE
(FUTURE THIRD VENTRICLE)

Cortical (parietal) NEP

Cortical (posterior cingulate) NEP

POSTERODORSAL POOL

Dorsal telencephalic choroid plexus

Dentate migration

Ammonic migration

Fornix

Thalamic NEP

Diencephalic choroid plexus

Future dorsal hippocampus

Fornical **GEP**

Cortical (dorsal hippocampal) NEP

TELENCEPHALIC SUPERVENTRICLE
(FUTURE LATERAL VENTRICLE)

MIDDORSAL POOL

Lateral migratory stream?

FUTURE BASAL GANGLIA

FUTURE PARACENTRAL LOBULE

Paracentral cortex

Paracentral *STF*

Cortical layers more prominent

FUTURE FRONTAL LOBE

Frontal cortex

Frontal *STF*

Interhemispheric fissure

Cingulate cortex

Cortical (anterior cingulate) NEP

ANTERODORSAL POOL

Cortical (paracentral) NEP

Cortical (frontal) NEP

Corticoganglionic NEP and SVZ

Anterolateral ganglionic NEP and SVZ

Arrows indicate the presumed *direction of neuron migration* from neuroepithelial sources.

62

PLATE 23A
CR 43 mm, GW10.6, C886
Horizontal/Frontal
Section 53

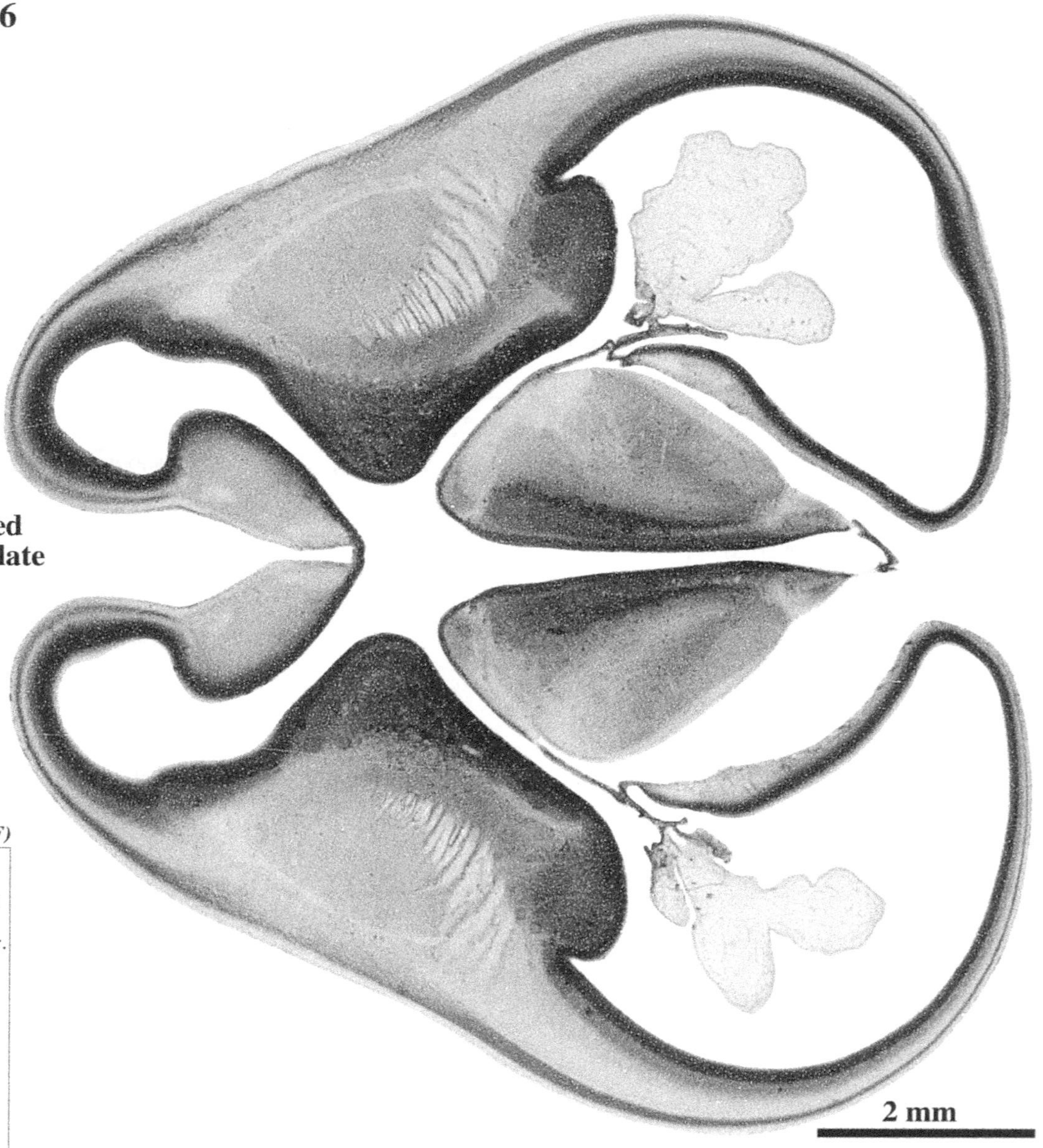

See the brain core enlarged in parts C and D of this plate on the following pages.

2 mm

LAYERS OF THE CORTICAL
STRATIFIED TRANSITIONAL FIELD (STF)

STF1 Superficial fibrous layer with an early developmental stage (*t1*) when many cells are migrating through it, followed by a late stage (*t2*) with sparse cells. Endures as the subcortical white matter.

STF4 Complex middle layer where sojourning and migrating cortical neurons grow corticofugal axons and intermingle with corticopetal axons.

STF5 Deep cellular layer that is prominent during the first trimester, the first sojourn zone to appear outside the germinal matrix.

PLATE 23B

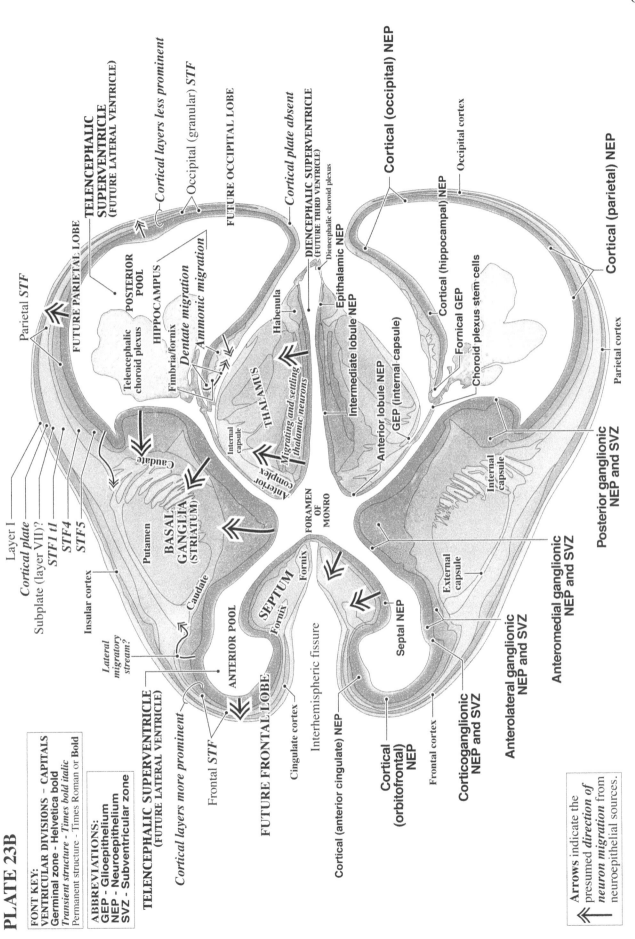

Arrows indicate the presumed *direction of neuron migration* from neuroepithelial sources.

64

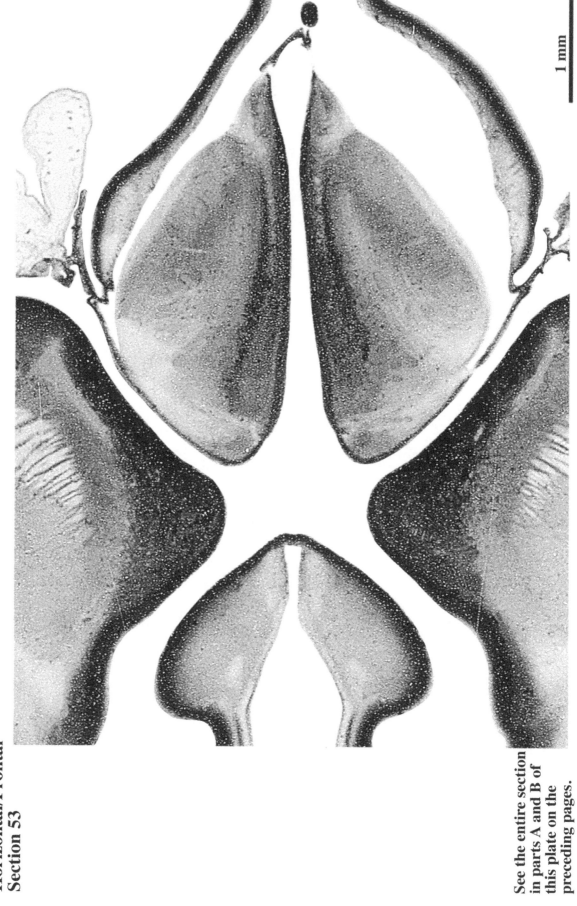

See the entire section in parts A and B of this plate on the preceding pages.

PLATE 23C
CR 43 mm, GW10.6, C886
Horizontal/Frontal
Section 53

1 mm

PLATE 23D

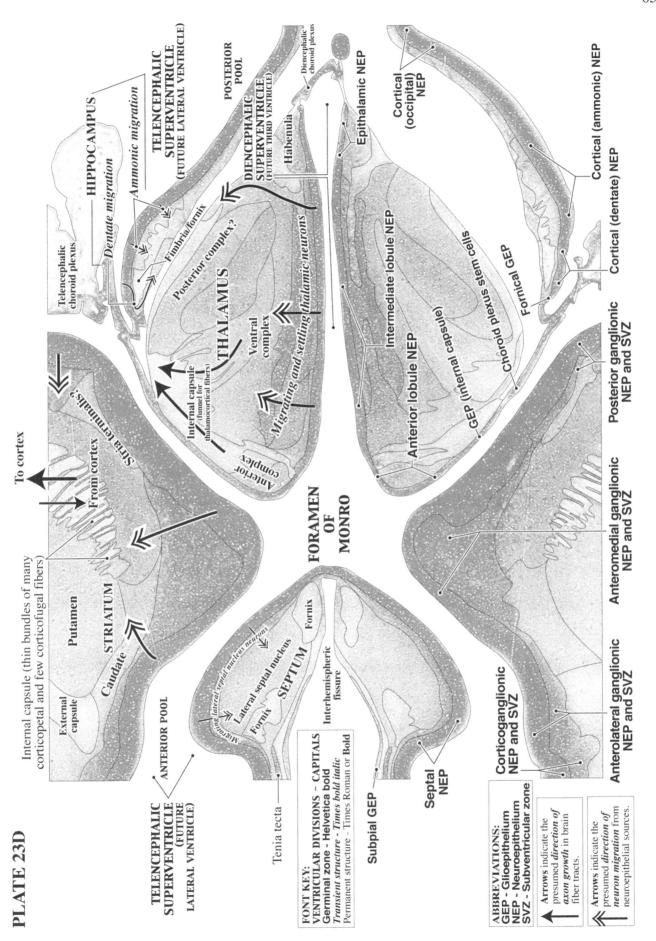

Internal capsule (thin bundles of many corticopetal and few corticofugal fibers)

To cortex

From cortex

Stria terminalis?

External capsule

Putamen

STRIATUM

Caudate

ANTERIOR POOL

TELENCEPHALIC SUPERVENTRICLE
(FUTURE LATERAL VENTRICLE)

Tenia tecta

Migrating lateral septal nucleus neurons

Lateral septal nucleus

SEPTUM

Fornix

Fornix

Interhemispheric fissure

FORAMEN OF MONRO

Subpial GEP

Septal NEP

Corticoganglionic NEP and SVZ

Anterolateral ganglionic NEP and SVZ

Anteromedial ganglionic NEP and SVZ

Posterior ganglionic NEP and SVZ

Anterior complex

Internal capsule (tunnel for thalamocortical fibers)

Migrating and settling thalamic neurons

Ventral complex

THALAMUS

Posterior complex?

Fimbria/fornix

Telencephalic choroid plexus

HIPPOCAMPUS

Dentate migration

Ammonic migration

TELENCEPHALIC SUPERVENTRICLE
(FUTURE LATERAL VENTRICLE)

POSTERIOR POOL

DIENCEPHALIC SUPERVENTRICLE
(FUTURE THIRD VENTRICLE)

Habenula

Diencephalic choroid plexus

Epithalamic NEP

Intermediate lobule NEP

Anterior lobule NEP

GEP (internal capsule)

Choroid plexus stem cells

Fornical GEP

Cortical (occipital) NEP

Cortical (ammonic) NEP

Cortical (dentate) NEP

FONT KEY:
VENTRICULAR DIVISIONS – CAPITALS
Germinal zone - Helvetica bold
Transient structure - Times bold italic
Permanent structure - Times Roman or **Bold**

ABBREVIATIONS:
GEP - Glioepithelium
NEP - Neuroepithelium
SVZ - Subventricular zone

Arrows indicate the presumed *direction of axon growth* in brain fiber tracts.

Arrows indicate the presumed *direction of neuron migration* from neuroepithelial sources.

66

PLATE 24A
CR 43 mm, GW10.6, C886
Horizontal/Frontal
Section 57

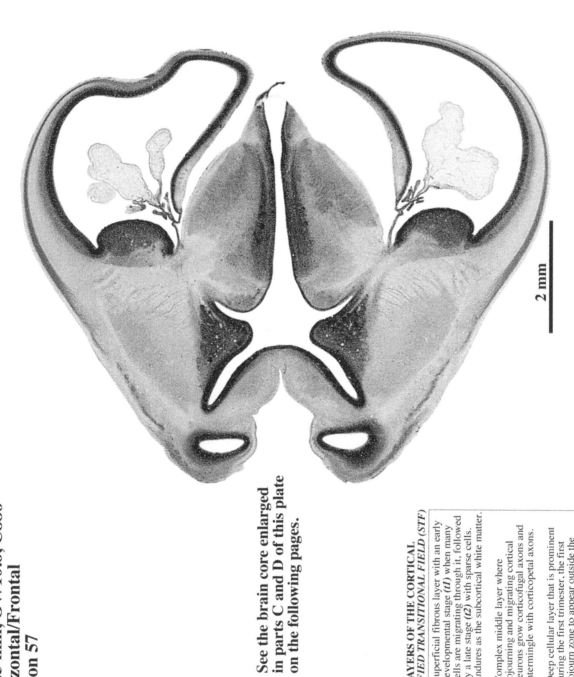

2 mm

See the brain core enlarged in parts C and D of this plate on the following pages.

LAYERS OF THE CORTICAL
STRATIFIED TRANSITIONAL FIELD (STF)

STF1 Superficial fibrous layer with an early developmental stage (*t1*) when many cells are migrating through it, followed by a late stage (*t2*) with sparse cells. Endures as the subcortical white matter.

STF4 Complex middle layer where sojourning and migrating cortical neurons grow corticofugal axons and intermingle with corticopetal axons.

STF5 Deep cellular layer that is prominent during the first trimester, the first sojourn zone to appear outside the germinal matrix.

PLATE 24B

FONT KEY:
VENTRICULAR DIVISIONS – CAPITALS
Germinal zone - Helvetica bold
Transient structure - Times bold italic
Permanent structure - Times Roman or **Bold**

ABBREVIATIONS:
GEP - Glioepithelium
NEP - Neuroepithelium
SVZ - Subventricular zone

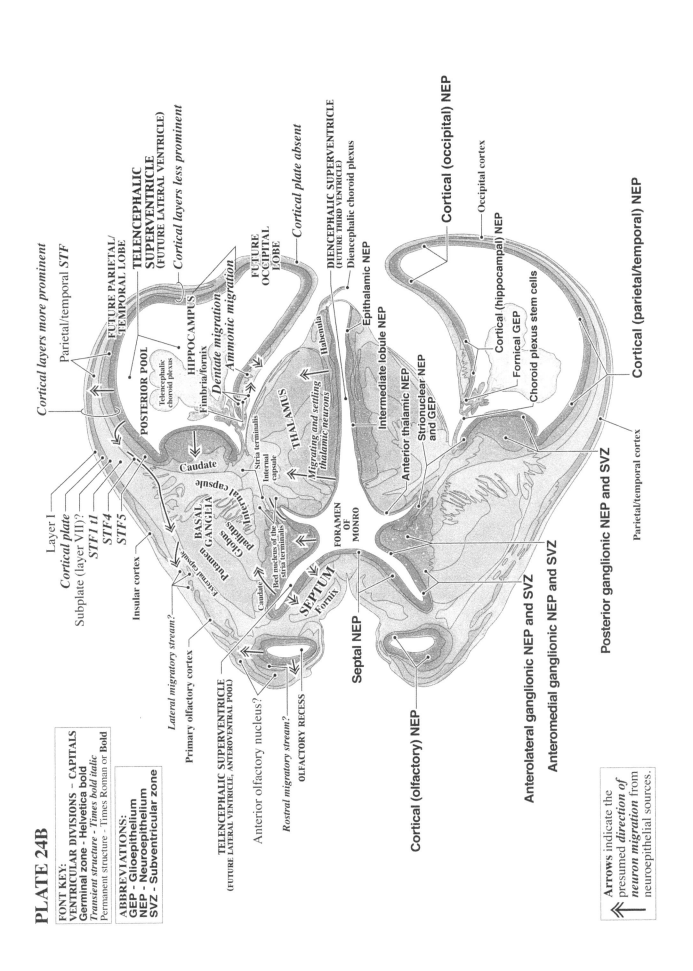

Cortical layers more prominent

Parietal/temporal *STF*

FUTURE PARIETAL/
TEMPORAL LOBE

**TELENCEPHALIC
SUPERVENTRICLE**
(FUTURE LATERAL VENTRICLE)

Cortical layers less prominent

HIPPOCAMPUS

POSTERIOR POOL

Telencephalic
choroid plexus

Fimbria/fornix

Dentate migration

Ammonic migration

*FUTURE
OCCIPITAL
LOBE*

Cortical plate absent

THALAMUS

Stria terminalis

Habenula

**DIENCEPHALIC
SUPERVENTRICLE**
(FUTURE THIRD VENTRICLE)

Diencephalic choroid plexus

Epithalamic NEP

*Migrating and settling
thalamic neurons*

Intermediate lobule NEP

Cortical (occipital) NEP

Occipital cortex

Cortical (parietal/temporal) NEP

Cortical (hippocampal) NEP

Fornical GEP

Choroid plexus stem cells

Anterior thalamic NEP

Strionuclear NEP
and GEP

Layer 1
Cortical plate
Subplate (layer VII)?
STF1 t1
STF4
STF5

Insular cortex

Caudate

Internal
capsule

*BASAL
GANGLIA*

*Globus
pallidus*

Putamen

Internal capsule

Foramen
of
MONRO

FORAMEN
OF
MONRO

Lateral migratory stream?

Primary olfactory cortex

External capsule

Bed nucleus of the
stria terminalis

Caudate

Parietal/temporal cortex

Posterior ganglionic NEP and SVZ

SEPTUM

Fornix

Septal NEP

Anterolateral ganglionic NEP and SVZ

Anteromedial ganglionic NEP and SVZ

TELENCEPHALIC SUPERVENTRICLE
(FUTURE LATERAL VENTRICLE, ANTEROVENTRAL POOL)

Anterior olfactory nucleus?

Rostral migratory stream?
OLFACTORY RECESS

Cortical (olfactory) NEP

Arrows indicate the
presumed *direction of
neuron migration* from
neuroepithelial sources.

68

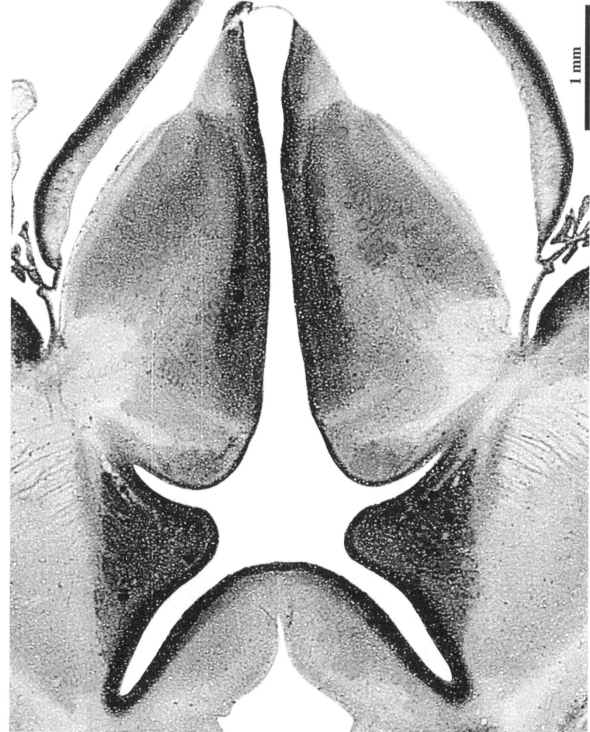

See the entire section in parts A and B of this plate on the preceding pages.

PLATE 24C
CR 43 mm,
GW10.6,
C886
Horizontal/Frontal
Section 57

1 mm

PLATE 24D

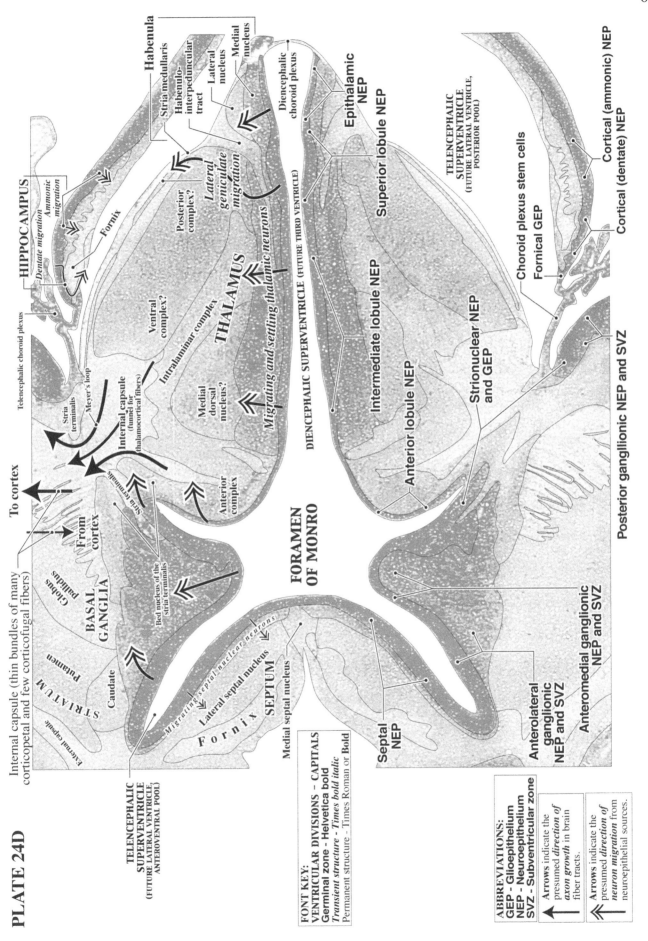

Internal capsule (thin bundles of many corticopetal and few corticofugal fibers)

To cortex

From cortex

HIPPOCAMPUS

Dentate migration

Ammonic migration

Fornix

Telencephalic choroid plexus

Habenula

Stria medullaris

Habenulo-interpeduncular tract

Lateral nucleus

Medial nucleus

Diencephalic choroid plexus

Epithalamic NEP

Superior lobule NEP

Posterior complex?

Lateral geniculate migration

Ventral complex?

THALAMUS

Intralaminar complex

Medial dorsal nucleus?

Migrating and settling thalamic neurons

DIENCEPHALIC SUPERVENTRICLE (FUTURE THIRD VENTRICLE)

Intermediate lobule NEP

Anterior lobule NEP

TELENCEPHALIC SUPERVENTRICLE (FUTURE LATERAL VENTRICLE, POSTERIOR POOL)

Choroid plexus stem cells

Fornical GEP

Cortical (ammonic) NEP

Cortical (dentate) NEP

Strionuclear NEP and GEP

Posterior gangllionic NEP and SVZ

Stria terminalis

Meyer's loop

Internal capsule (funnel for thalamocortical fibers)

Anterior complex

stria terminalis

Bed nucleus of the stria terminalis

TELENCEPHALIC SUPERVENTRICLE (FUTURE LATERAL VENTRICLE, ANTEROVENTRAL POOL)

Globus pallidus

BASAL GANGLIA

External capsule

Putamen

Caudate

STRIATUM

Migrating septal nuclear neurons

Lateral septal nucleus

SEPTUM

Fornix

Medial septal nucleus

FORAMEN OF MONRO

Septal NEP

Anterolateral gangllionic NEP and SVZ

Anteromedial gangllionic NEP and SVZ

FONT KEY:
VENTRICULAR DIVISIONS – CAPITALS
Germinal zone – **Helvetica bold**
Transient structure – Times bold italic
Permanent structure – Times Roman or **Bold**

ABBREVIATIONS:
GEP – Glioepithelium
NEP – Neuroepithelium
SVZ – Subventricular zone

Arrows indicate the presumed *direction of axon growth* in brain fiber tracts.

Arrows indicate the presumed *direction of neuron migration* from neuroepithelial sources.

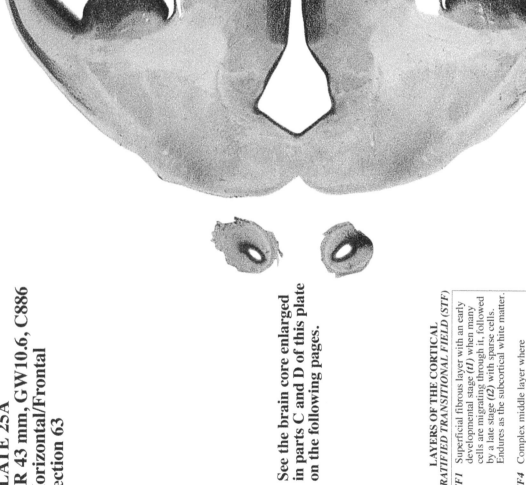

PLATE 25A
CR 43 mm, GW10.6, C886
Horizontal/Frontal
Section 63

See the brain core enlarged
in parts C and D of this plate
on the following pages.

2 mm

LAYERS OF THE CORTICAL
STRATIFIED TRANSITIONAL FIELD (STF)

STF1 Superficial fibrous layer with an early
developmental stage (*t1*) when many
cells are migrating through it, followed
by a late stage (*t2*) with sparse cells.
Endures as the subcortical white matter.

STF4 Complex middle layer where
sojourning and migrating cortical
neurons grow corticofugal axons and
intermingle with corticopetal axons.

STF5 Deep cellular layer that is prominent
during the first trimester, the first
sojourn zone to appear outside the
germinal matrix.

71

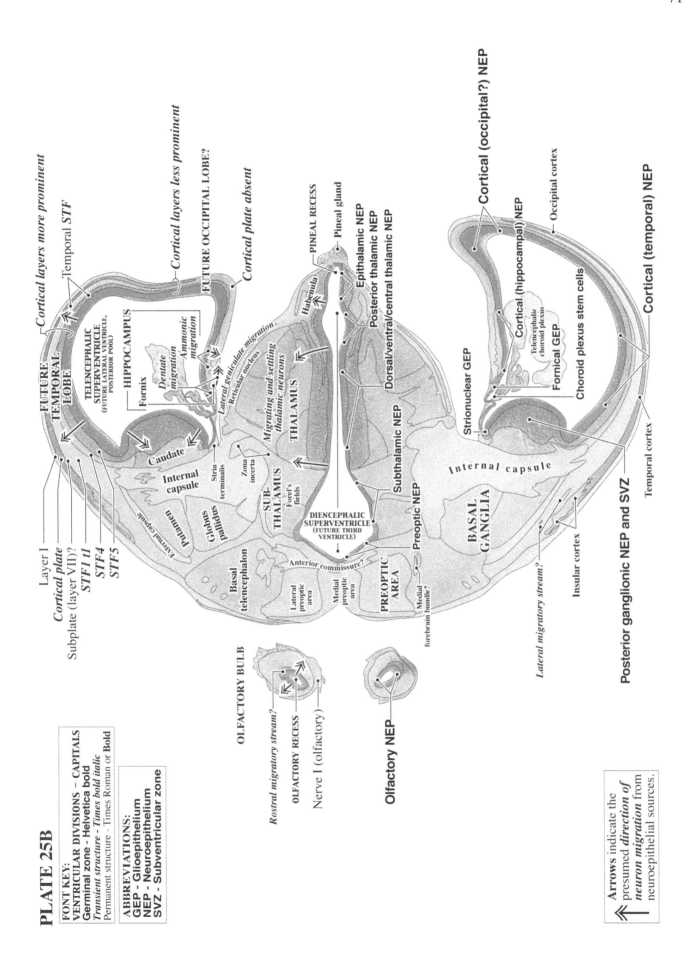

PLATE 25B

FONT KEY:
VENTRICULAR DIVISIONS – CAPITALS
Germinal zone - **Helvetica bold**
Transient structure - Times bold italic
Permanent structure - Times Roman or **Bold**

ABBREVIATIONS:
GEP - Glioepithelium
NEP - Neuroepithelium
SVZ - Subventricular zone

Arrows indicate the presumed *direction of neuron migration* from neuroepithelial sources.

72

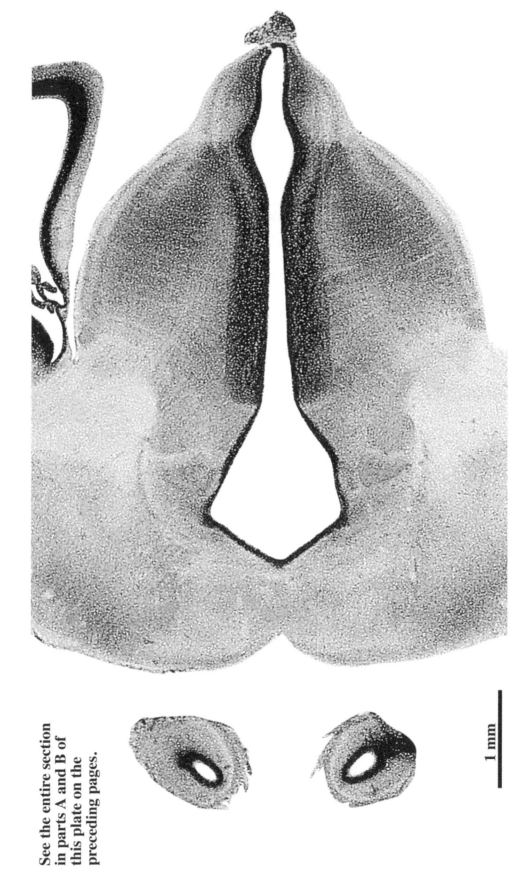

PLATE 25C
CR 43 mm, GW10.6, C886
Horizontal/Frontal
Section 63

See the entire section
in parts A and B of
this plate on the
preceding pages.

1 mm

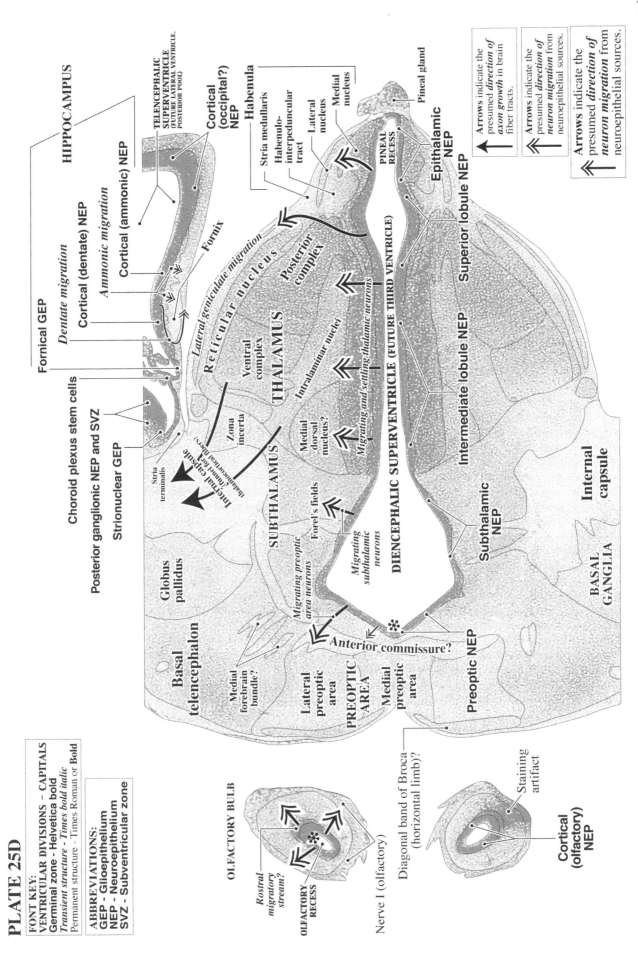

PLATE 25D

FONT KEY:
VENTRICULAR DIVISIONS - CAPITALS
Germinal zone - Helvetica bold
Transient structure - Times bold italic
Permanent structure - Times Roman or Bold

ABBREVIATIONS:
GEP - Glioepithelium
NEP - Neuroepithelium
SVZ - Subventricular zone

73

HIPPOCAMPUS

Fornical GEP

TELENCEPHALIC SUPERVENTRICLE (FUTURE LATERAL VENTRICLE, POSTERIOR POOL)

Cortical (occipital?) NEP

Habenula

Stria medullaris

Habenulo-interpeduncular tract

Lateral nucleus

Medial nucleus

Pineal gland

Arrows indicate the presumed *direction of axon growth* in brain fiber tracts.

Arrows indicate the presumed *direction of neuron migration* from neuroepithelial sources.

Arrows indicate the presumed *direction of neuron migration* from neuroepithelial sources.

Dentate migration

Cortical (dentate) NEP

Ammonic migration

Cortical (ammonic) NEP

Fornix

Choroid plexus stem cells

Posterior ganglionic NEP and SVZ

Strionuclear GEP

Lateral geniculate migration

Reticular nucleus

Posterior complex

THALAMUS

Ventral complex

Intralaminar nuclei

Migrating and settling thalamic neurons

PINEAL RECESS

Epithalamic NEP

Superior lobule NEP

Zona incerta

Stria terminalis

Medial dorsal nucleus?

Internal capsule (band for) thalamocortical fibers

SUBTHALAMUS

Forel's fields

DIENCEPHALIC SUPERVENTRICLE (FUTURE THIRD VENTRICLE)

Intermediate lobule NEP

Subthalamic NEP

INTERNAL capsule

Globus pallidus

Migrating subthalamic neurons

Basal telencephalon

Migrating preoptic area neurons

Medial forebrain bundle?

Anterior commissure?

Lateral preoptic area

PREOPTIC AREA

Medial preoptic area

Preoptic NEP

BASAL GANGLIA

OLFACTORY BULB

Rostral migratory stream?

OLFACTORY RECESS

Nerve I (olfactory)

Diagonal band of Broca (horizontal limb)?

Staining artifact

Cortical (olfactory) NEP

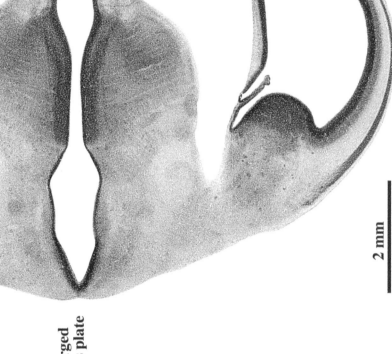

PLATE 26A
CR 43 mm, GW10.6, C886
Horizontal/Frontal
Section 69

See the brain core enlarged in parts C and D of this plate on the following pages.

2 mm

LAYERS OF THE CORTICAL
STRATIFIED TRANSITIONAL FIELD (STF)

STF1   Superficial fibrous layer with an early developmental stage (*t1*) when many cells are migrating through it, followed by a late stage (*t2*) with sparse cells. Endures as the subcortical white matter.

STF4   Complex middle layer where sojourning and migrating cortical neurons grow corticofugal axons and intermingle with corticopetal axons.

STF5   Deep cellular layer that is prominent during the first trimester, the first sojourn zone to appear outside the germinal matrix.

PLATE 26B

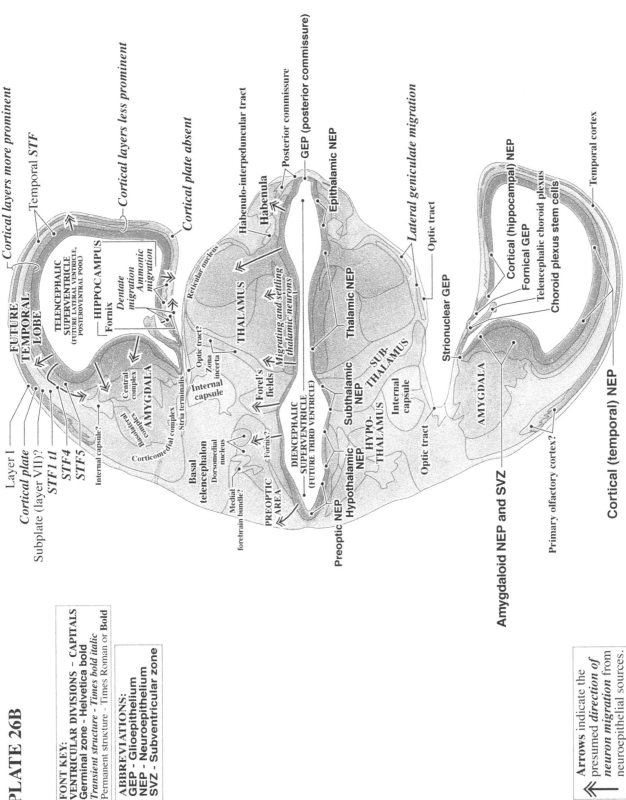

Arrows indicate the
presumed *direction of
neuron migration* from
neuroepithelial sources.

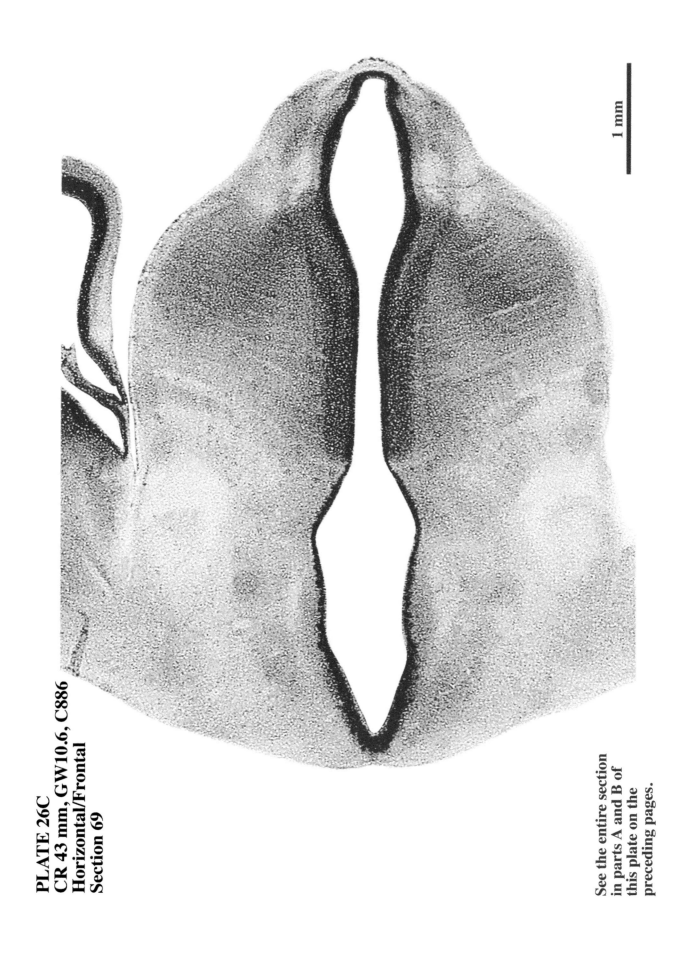

1 mm

PLATE 26C
CR 43 mm, GW10.6, C886
Horizontal/Frontal
Section 69

See the entire section
in parts A and B of
this plate on the
preceding pages.

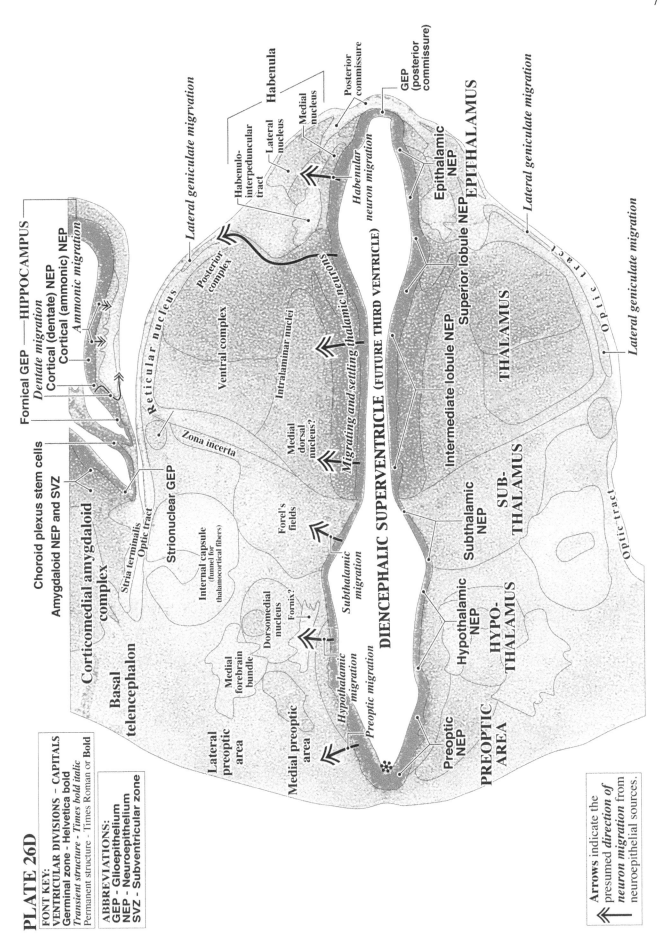

PLATE 26D

FONT KEY:
VENTRICULAR DIVISIONS – CAPITALS
Germinal zone – Helvetica bold
Transient structure – Times bold italic
Permanent structure – Times Roman or Bold

ABBREVIATIONS:
GEP – Glioepithelium
NEP – Neuroepithelium
SVZ – Subventricular zone

Fornical GEP ⎯⎯ **HIPPOCAMPUS**
Dentate migration
Cortical (dentate) NEP
Cortical (ammonic) NEP
Ammonic migration

Choroid plexus stem cells

Amygdaloid NEP and SVZ

Corticomedial amygdaloid complex

Basal telencephalon

Stria terminalis
Optic tract

Strionuclear GEP

Lateral preoptic area

Internal capsule
(funnel for thalamocortical fibers)

Medial forebrain bundle

Dorsomedial nucleus
Fornix?

Medial preoptic area

Hypothalamic migration

Preoptic migration

Lateral geniculate migrvation

Ammonic migration

Posterior complex

R e t i c u l a r   n u c l e u s

Zona incerta

Ventral complex

Intralaminar nuclei

Medial dorsal nucleus?

Migrating and settling thalamic neurons

Forel's fields

Subthalamic migration

Habenulo-interpeduncular tract

Lateral nucleus

Medial nucleus

Habenula

Habenular neuron migration

Posterior commissure

GEP (posterior commissure)

DIENCEPHALIC SUPERVENTRICLE (FUTURE THIRD VENTRICLE)

Epithalamic NEP

EPITHALAMUS

Superior lobule NEP

Intermediate lobule NEP

THALAMUS

Subthalamic NEP

SUB-THALAMUS

Hypothalamic NEP

HYPO-THALAMUS

Preoptic NEP

PREOPTIC AREA

Lateral geniculate migration

Lateral geniculate migration

O p t i c   t r a c t

Lateral geniculate migration

Optic tract

Arrows indicate the presumed *direction of neuron migration* from neuroepithelial sources.

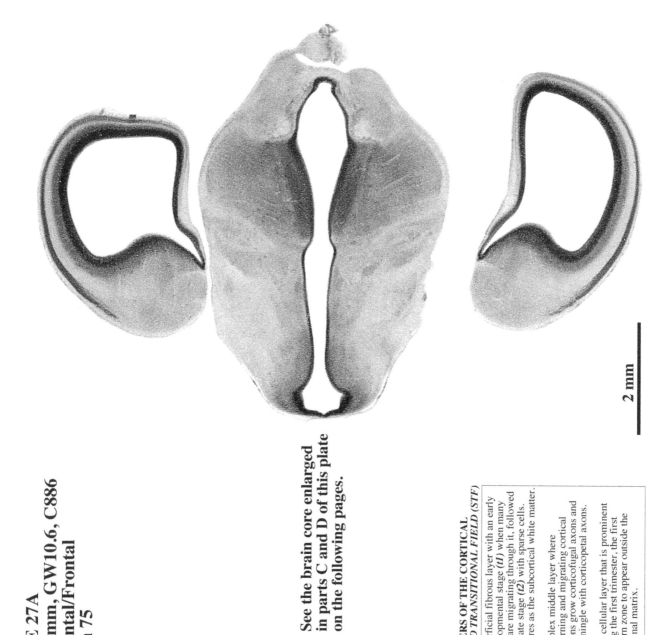

PLATE 27A
CR 43 mm, GW10.6, C886
Horizontal/Frontal
Section 75

See the brain core enlarged in parts C and D of this plate on the following pages.

LAYERS OF THE CORTICAL
STRATIFIED TRANSITIONAL FIELD (STF)

STF1   Superficial fibrous layer with an early developmental stage (*t1*) when many cells are migrating through it, followed by a late stage (*t2*) with sparse cells. Endures as the subcortical white matter.

STF4   Complex middle layer where sojourning and migrating cortical neurons grow corticofugal axons and intermingle with corticopetal axons.

STF5   Deep cellular layer that is prominent during the first trimester, the first sojourn zone to appear outside the germinal matrix.

2 mm

PLATE 27B

FONT KEY:
VENTRICULAR DIVISIONS - CAPITALS
Germinal zone - Helvetica bold
Transient structure - Times bold italic
Permanent structure - Times Roman or **Bold**

ABBREVIATIONS:
GEP - Glioepithelium
NEP - Neuroepithelium
SVZ - Subventricular zone

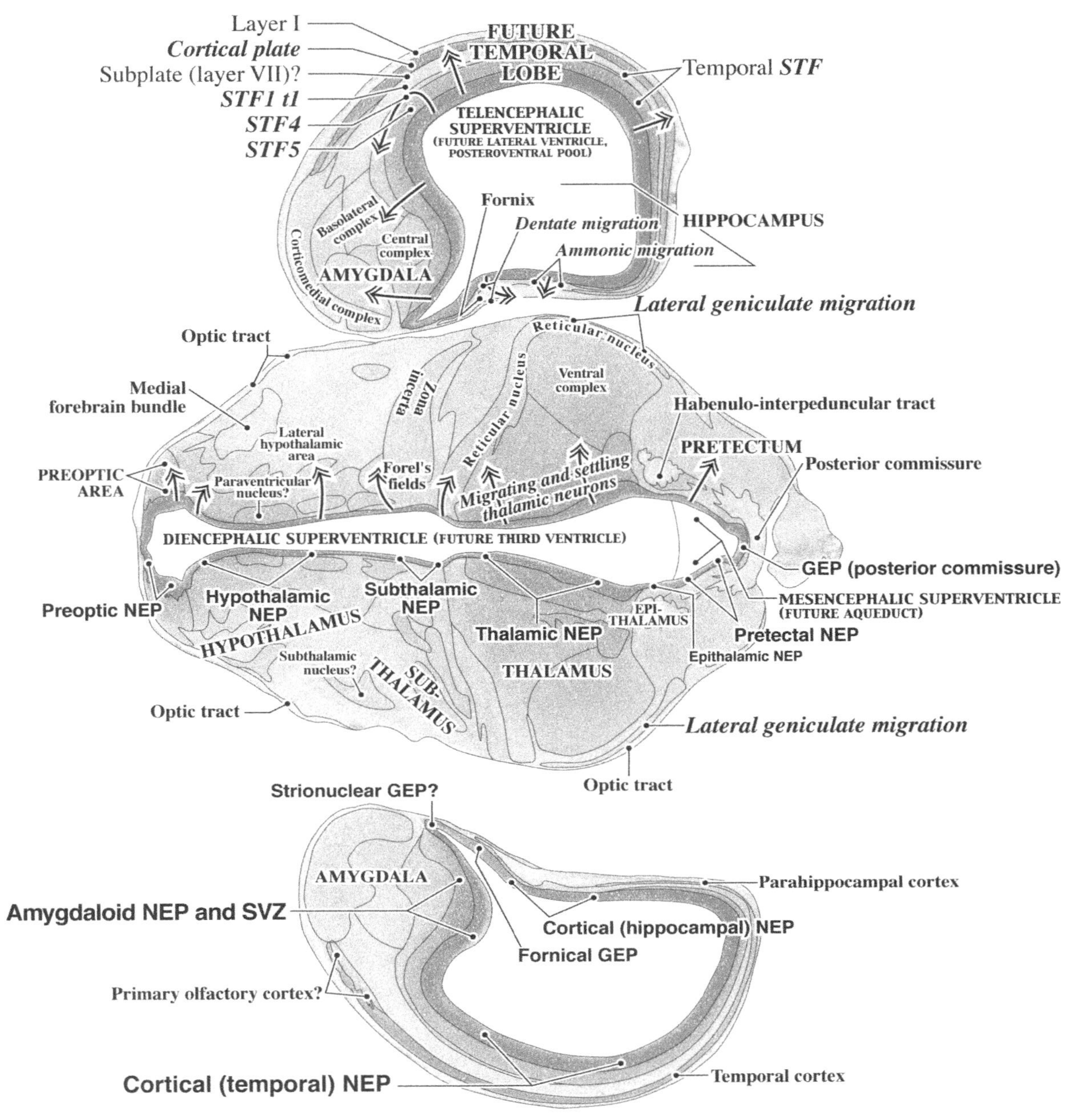

Layer 1
Cortical plate
Subplate (layer VII)?
STF1 t1
STF4
STF5

Temporal *STF*

FUTURE
TEMPORAL
LOBE

HIPPOCAMPUS

TELENCEPHALIC
SUPERVENTRICLE
(FUTURE LATERAL VENTRICLE,
POSTEROVENTRAL POOL)

Dentate migration
Ammonic migration

Fornix

Lateral geniculate migration

Central
complex

Basolateral
complex

AMYGDALA

Corticomedial complex

Optic tract

Medial
forebrain
bundle

PREOPTIC
AREA

Preoptic NEP

Lateral
hypothalamic
area

Paraventricular
nucleus?

DIENCEPHALIC SUPERVENTRICLE (FUTURE THIRD VENTRICLE)

Hypothalamic
NEP
HYPOTHALAMUS

Subthalamic
nucleus?

SUB-
THALAMUS

Optic tract

Zona
incerta

Forel's
fields

*Migrating and settling
thalamic neurons*

Reticular nucleus

Subthalamic
NEP

Thalamic NEP

THALAMUS

EPI-
THALAMUS

Epithalamic NEP

Reticular nucleus

Ventral
complex

Habenulo-interpeduncular tract

PRETECTUM

Posterior commissure

GEP (posterior commissure)

MESENCEPHALIC SUPERVENTRICLE
(FUTURE AQUEDUCT)

Pretectal NEP

Lateral geniculate migration

Optic tract

Strionuclear GEP?

Amygdaloid NEP and SVZ

AMYGDALA

Parahippocampal cortex

Cortical (hippocampal) NEP

Fornical GEP

Temporal cortex

Primary olfactory cortex?

Cortical (temporal) NEP

Arrows indicate the
presumed *direction of
neuron migration* from
neuroepithelial sources.

80

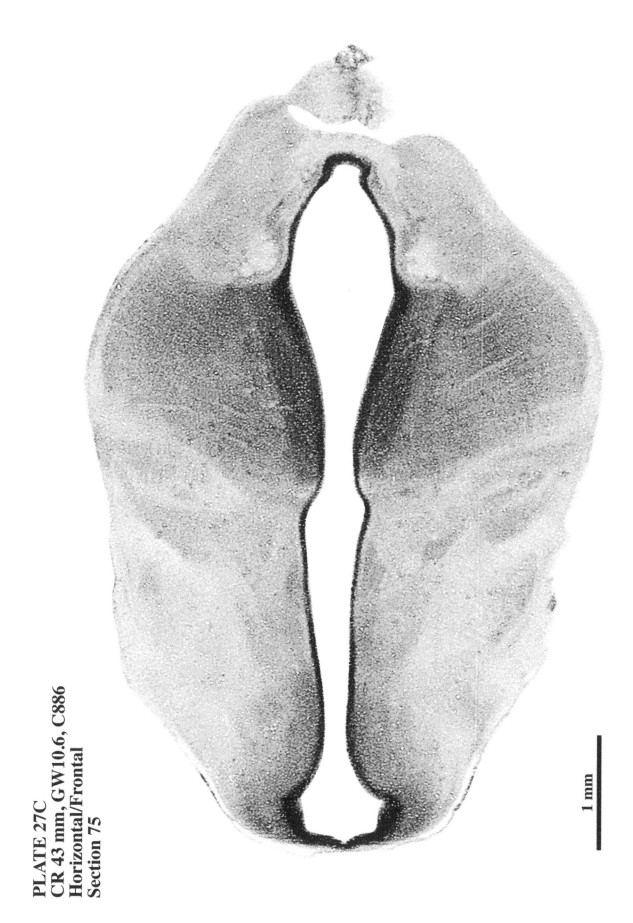

1 mm

See the entire section in parts A and B of this plate on the preceding pages.

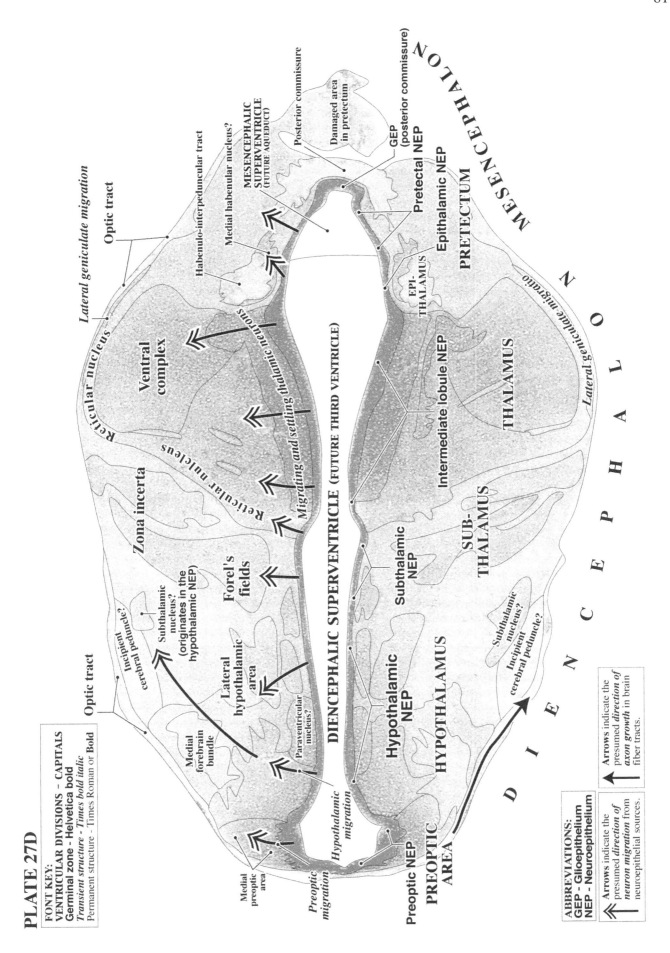

PLATE 27D

FONT KEY:
VENTRICULAR DIVISIONS – CAPITALS
Germinal zone – Helvetica bold
Transient structure - Times bold italic
Permanent structure - Times Roman or **Bold**

ABBREVIATIONS:
GEP - Glioepithelium
NEP - Neuroepithelium

Arrows indicate the presumed *direction of neuron migration* from neuroepithelial sources.

Arrows indicate the presumed *direction of axon growth* in brain fiber tracts.

Posterior commissure

Damaged area in pretectum

GEP (posterior commissure)

Pretectal NEP

Medial habenular nucleus?

Habenulo-interpeduncular tract

MESENCEPHALIC SUPERVENTRICLE (FUTURE AQUEDUCT)

Epithalamic NEP

EPI-THALAMUS

PRETECTUM

Lateral geniculate migration

Optic tract

Reticular nucleus?

Ventral complex

Migrating and settling thalamic neurons

Zona incerta

Reticular nuclei

Intermediate lobule NEP

THALAMUS

Forel's fields

Subthalamic nucleus? (originates in the hypothalamic NEP)

Lateral hypothalamic area

Paraventricular nucleus?

Incipient cerebral peduncle?

Subthalamic NEP

SUB-THALAMUS

Subthalamic nucleus? Incipient cerebral peduncle?

DIENCEPHALIC SUPERVENTRICLE (FUTURE THIRD VENTRICLE)

Optic tract

Medial forebrain bundle

Hypothalamic migration

Hypothalamic NEP

HYPOTHALAMUS

Medial preoptic area

Preoptic migration

Preoptic NEP

PREOPTIC AREA

M E S E N C E P H A L O N

Lateral geniculate migratio

D I E N C E P H A L O N

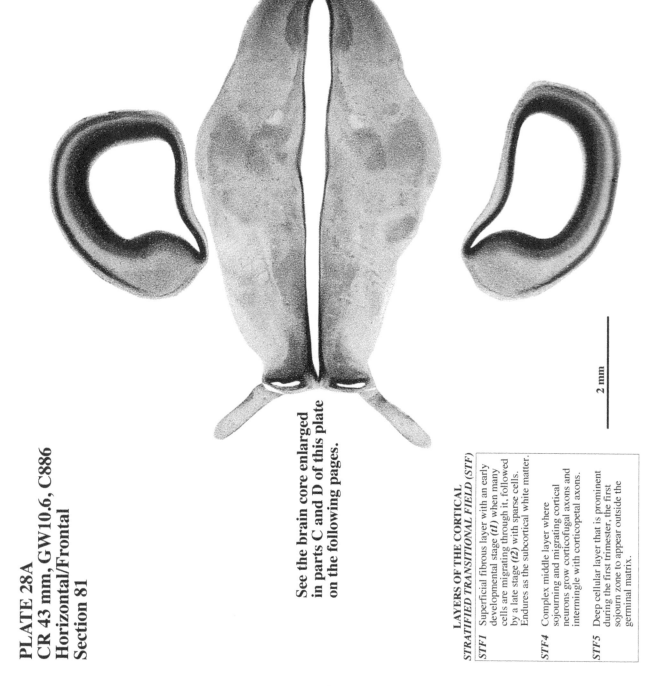

See the brain core enlarged in parts C and D of this plate on the following pages.

PLATE 28A
CR 43 mm, GW10.6, C886
Horizontal/Frontal
Section 81

2 mm

LAYERS OF THE CORTICAL
STRATIFIED TRANSITIONAL FIELD (STF)

STF1 Superficial fibrous layer with an early developmental stage (*t1*) when many cells are migrating through it, followed by a late stage (*t2*) with sparse cells. Endures as the subcortical white matter.

STF4 Complex middle layer where sojourning and migrating cortical neurons grow corticofugal axons and intermingle with corticopetal axons.

STF5 Deep cellular layer that is prominent during the first trimester, the first sojourn zone to appear outside the germinal matrix.

PLATE 28B

FONT KEY:
VENTRICULAR DIVISIONS - CAPITALS
Germinal zone - Helvetica bold
Transient structure - Times bold italic
Permanent structure - Times Roman or **Bold**

ABBREVIATIONS:
GEP - Glioepithelium
NEP - Neuroepithelium
SVZ - Subventricular zone

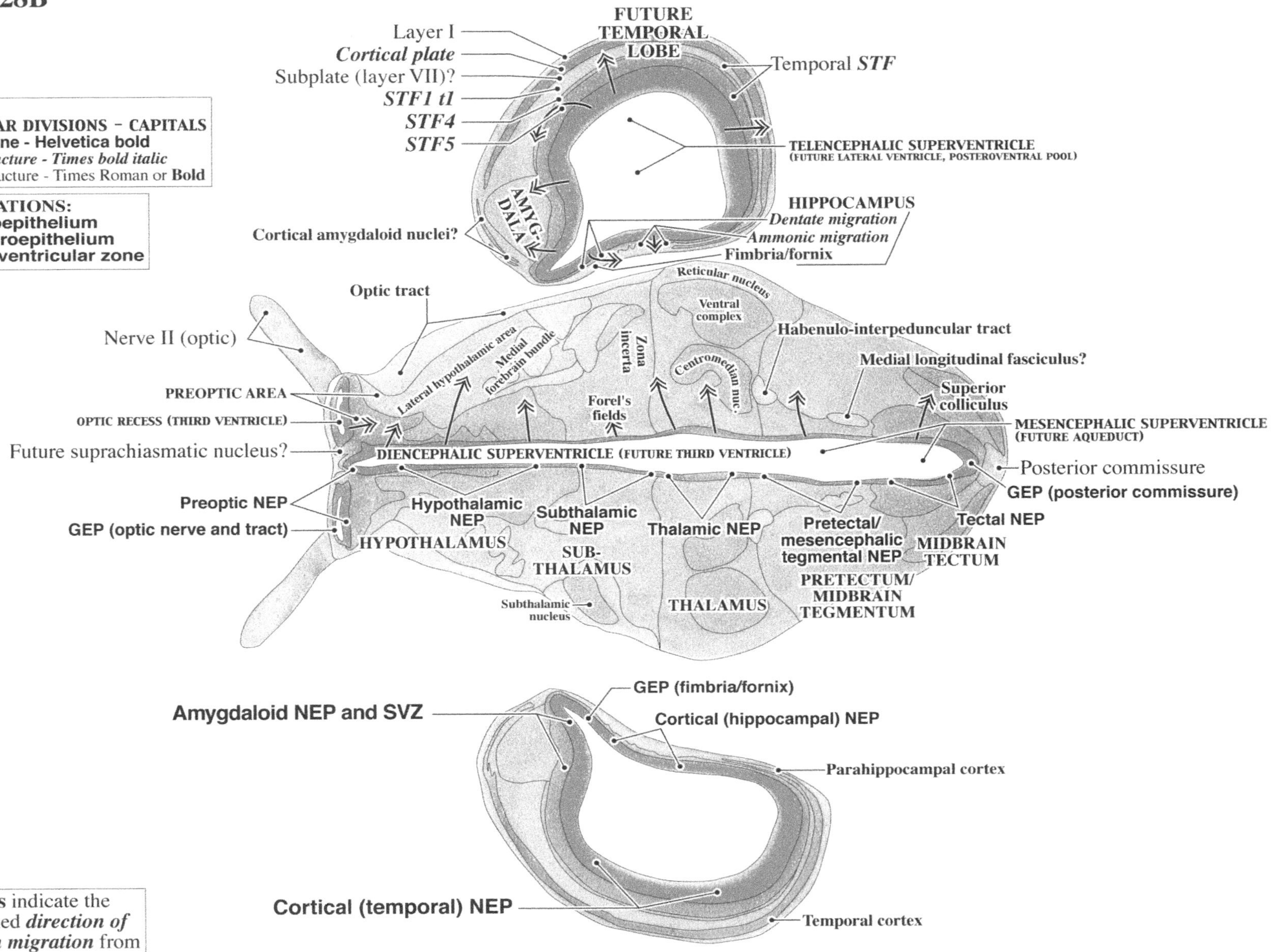

FUTURE TEMPORAL LOBE
Layer I
Cortical plate
Subplate (layer VII)?
STF1 t1
STF4
STF5
Temporal STF
TELENCEPHALIC SUPERVENTRICLE
(FUTURE LATERAL VENTRICLE, POSTEROVENTRAL POOL)
AMYG-DALA
Cortical amygdaloid nuclei?
HIPPOCAMPUS
Dentate migration
Ammonic migration
Fimbria/fornix
Optic tract
Nerve II (optic)
Reticular nucleus
Ventral complex
Zona incerta
Lateral hypothalamic area
Medial forebrain bundle
Centromedian nuc.
Habenulo-interpeduncular tract
Medial longitudinal fasciculus?
Superior colliculus
PREOPTIC AREA
OPTIC RECESS (THIRD VENTRICLE)
Forel's fields
MESENCEPHALIC SUPERVENTRICLE
(FUTURE AQUEDUCT)
Future suprachiasmatic nucleus?
DIENCEPHALIC SUPERVENTRICLE (FUTURE THIRD VENTRICLE)
Posterior commissure
GEP (posterior commissure)
Preoptic NEP
Hypothalamic NEP
Subthalamic NEP
Thalamic NEP
Pretectal/ mesencephalic tegmental NEP
Tectal NEP
GEP (optic nerve and tract)
HYPOTHALAMUS
SUB-THALAMUS
MIDBRAIN TECTUM
PRETECTUM/ MIDBRAIN TEGMENTUM
Subthalamic nucleus
THALAMUS
GEP (fimbria/fornix)
Amygdaloid NEP and SVZ
Cortical (hippocampal) NEP
Parahippocampal cortex
Cortical (temporal) NEP
Temporal cortex

Arrows indicate the presumed ***direction of neuron migration*** from neuroepithelial sources.

PLATE 28C
CR 43 mm, GW10.6, C886
Horizontal/Frontal
Section 81

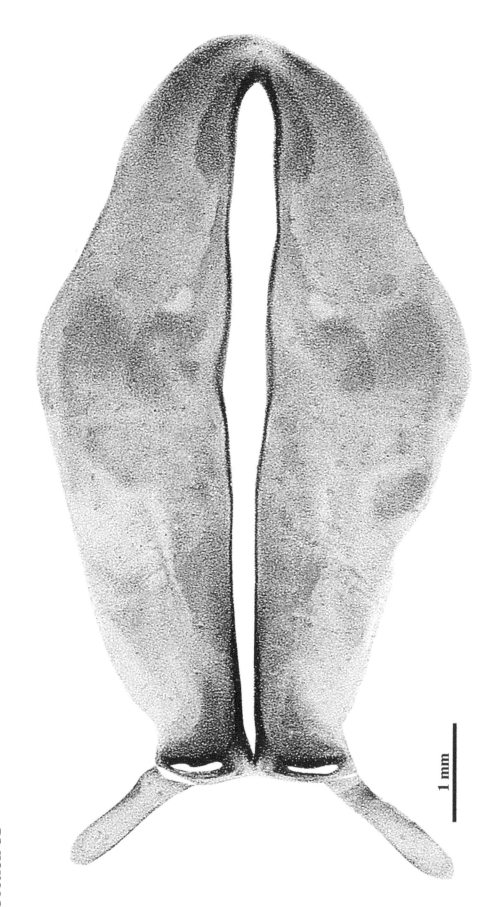

1 mm

See the entire section in parts A and B of this plate on the preceding pages.

PLATE 28D

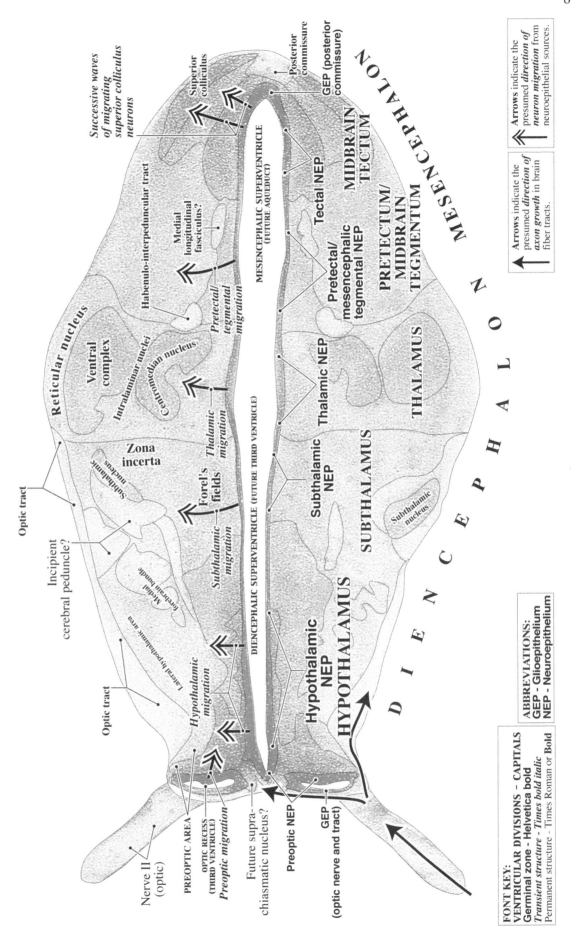

85

Successive waves of migrating superior colliculus neurons

Superior colliculus

Posterior commissure

GEP (posterior commissure)

MIDBRAIN TECTUM

Tectal NEP

MESENCEPHALIC SUPERVENTRICLE (FUTURE AQUEDUCT)

Habenulo-interpeduncular tract

Medial longitudinal fasciculus?

Pretectal/ tegmental migration

Pretectal/ mesencephalic tegmental NEP

PRETECTUM/ MIDBRAIN TEGMENTUM

Reticular nucleus

Ventral complex

Intralaminar nuclei

Centromedian nucleus

Thalamic migration

Thalamic NEP

THALAMUS

MESENCEPHALON

Zona incerta

Subthalamic nucleus

Forel's fields

Subthalamic migration

DIENCEPHALIC SUPERVENTRICLE (FUTURE THIRD VENTRICLE)

Subthalamic NEP

SUBTHALAMUS

Subthalamic nucleus

Optic tract

Incipient cerebral peduncle?

Medial forebrain bundle

Optic tract

Lateral hypothalamic area

Hypothalamic migration

Hypothalamic NEP

HYPOTHALAMUS

DIENCEPHALON

Nerve II (optic)

PREOPTIC AREA

OPTIC RECESS (THIRD VENTRICLE)
Preoptic migration

Future supra- chiasmatic nucleus?

Preoptic NEP

GEP (optic nerve and tract)

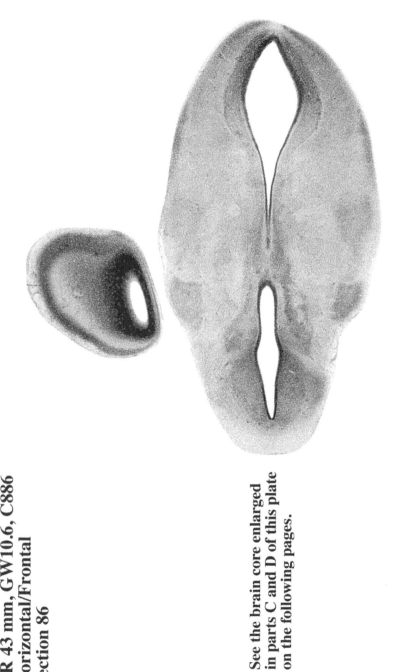

2 mm

PLATE 29A
CR 43 mm, GW10.6, C886
Horizontal/Frontal
Section 86

See the brain core enlarged in parts C and D of this plate on the following pages.

LAYERS OF THE CORTICAL
STRATIFIED TRANSITIONAL FIELD (STF)

STF1 Superficial fibrous layer with an early developmental stage (*t1*) when many cells are migrating through it, followed by a late stage (*t2*) with sparse cells. Endures as the subcortical white matter.

STF4 Complex middle layer where sojourning and migrating cortical neurons grow corticofugal axons and intermingle with corticopetal axons.

STF5 Deep cellular layer that is prominent during the first trimester, the first sojourn zone to appear outside the germinal matrix.

PLATE 29B

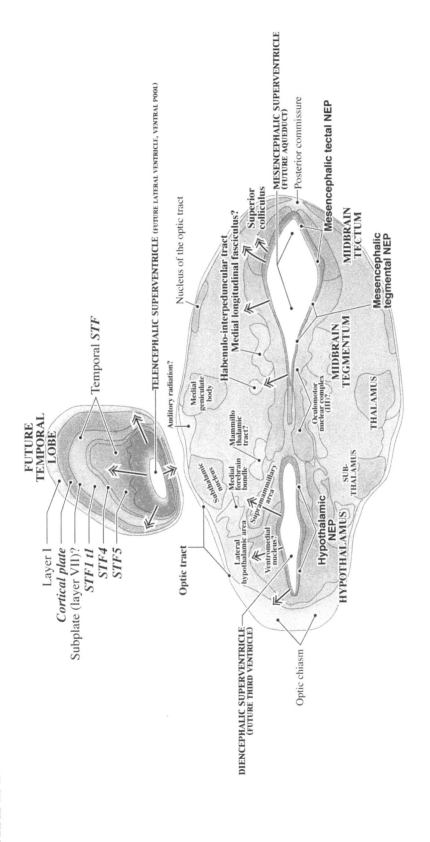

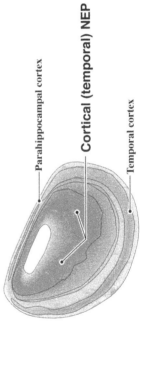

FUTURE
TEMPORAL
LOBE

Temporal *STF*

Layer 1
Cortical plate
Subplate (layer VII)?
STF1 t1
STF4
STF5

TELENCEPHALIC SUPERVENTRICLE (FUTURE LATERAL VENTRICLE, VENTRAL POOL)

Nucleus of the optic tract

Auditory radiation?

Medial
geniculate
body

Mammillo
thalamic
tract?

Optic tract

Subthalamic
nucleus

Medial
forebrain
bundle

Supramammillary
area

Lateral
hypothalamic area

Ventromedial
nucleus?

**Hypothalamic
NEP**

HYPOTHALAMUS

Optic chiasm

DIENCEPHALIC SUPERVENTRICLE
(FUTURE THIRD VENTRICLE)

Habenulo-interpeduncular tract
Medial longitudinal fasciculus?

**Superior
colliculus**

MESENCEPHALIC SUPERVENTRICLE
(FUTURE AQUEDUCT)

Posterior commissure

Mesencephalic tectal NEP

**MIDBRAIN
TECTUM**

**Mesencephalic
tegmental NEP**

Oculomotor
nuclear complex
(III)?

**MIDBRAIN
TEGMENTUM**

THALAMUS

**SUB-
THALAMUS**

Parahippocampal cortex

Cortical (temporal) NEP

Temporal cortex

NEP - Neuroepithelium

⇐ **Arrows** indicate the
presumed *direction of
neuron migration* from
neuroepithelial sources.

FONT KEY:
VENTRICULAR DIVISIONS – CAPITALS
Germinal zone - **Helvetica bold**
Transient structure - *Times bold italic*
Permanent structure - Times Roman or **Bold**

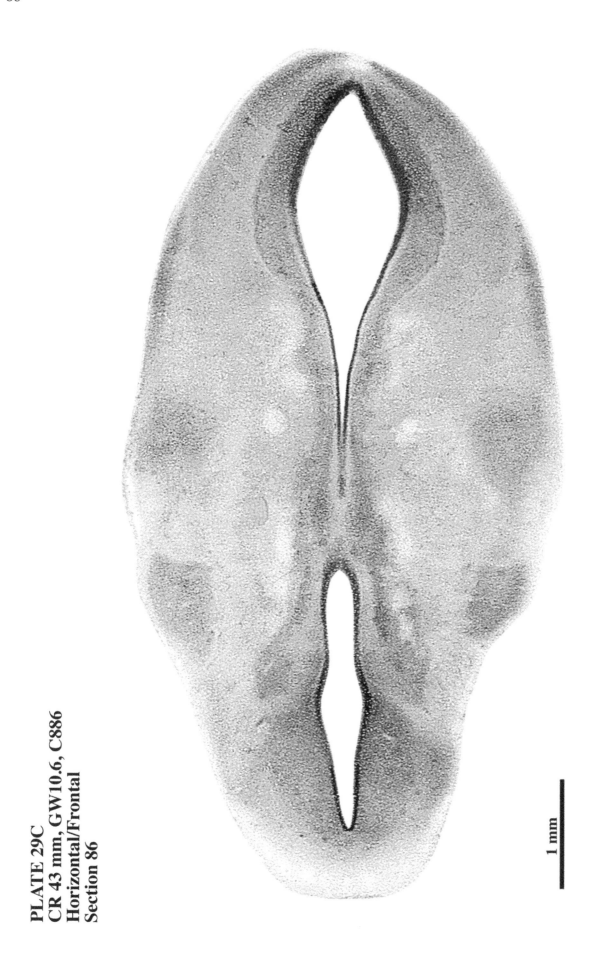

PLATE 29C
CR 43 mm, GW10.6, C886
Horizontal/Frontal
Section 86

1 mm

See the entire section in parts A and B of this plate on the preceding pages.

PLATE 29D

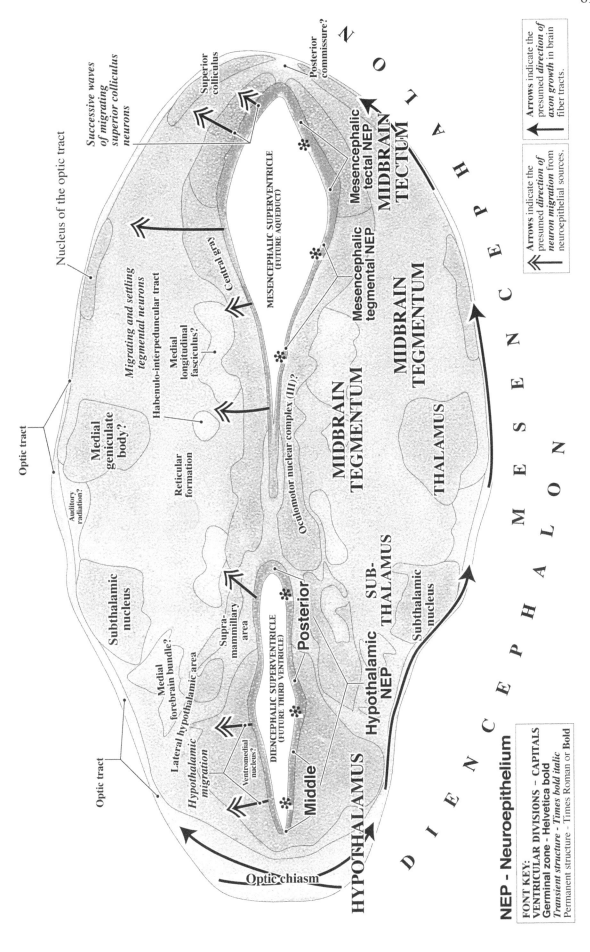

Posterior

Middle

Nucleus of the optic tract

Successive waves of migrating superior colliculus neurons

Superior colliculus

Posterior commissure?

Optic tract

Migrating and settling tegmental neurons

Central gray

Habenulo-interpeduncular tract

MESENCEPHALIC SUPERVENTRICLE (FUTURE AQUEDUCT)

Mesencephalic tectal NEP

MIDBRAIN TECTUM

Medial geniculate body?

Medial longitudinal fasciculus?

Mesencephalic tegmental NEP

Auditory radiation?

Optic tract

Subthalamic nucleus

Reticular formation

Oculomotor nuclear complex (III)?

MIDBRAIN TEGMENTUM

MIDBRAIN TEGMENTUM

THALAMUS

Supra-mammillary area

SUB-THALAMUS

Subthalamic nucleus

Medial forebrain bundle?

Lateral hypothalamic area

DIENCEPHALIC SUPERVENTRICLE (FUTURE THIRD VENTRICLE)

Hypothalamic NEP

Hypothalamic migration

Ventromedial nucleus?

Optic tract

HYPOTHALAMUS

Optic chiasm

M E S E N C E P H A L O N

D I E N C E P H A L O N

NEP - Neuroepithelium

FONT KEY:
VENTRICULAR DIVISIONS – CAPITALS
Germinal zone - Helvetica bold
Transient structure - Times bold italic
Permanent structure - Times Roman or **Bold**

Arrows indicate the presumed *direction of neuron migration* from neuroepithelial sources.

Arrows indicate the presumed *direction of axon growth* in brain fiber tracts.

**PLATE 30A
CR 43 mm, GW10.6, C886
Horizontal/Frontal
Section 90**

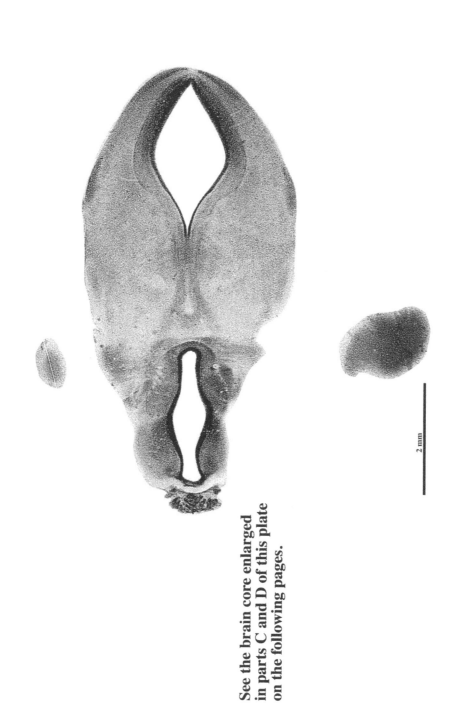

See the brain core enlarged
in parts C and D of this plate
on the following pages.

2 mm

PLATE 30B

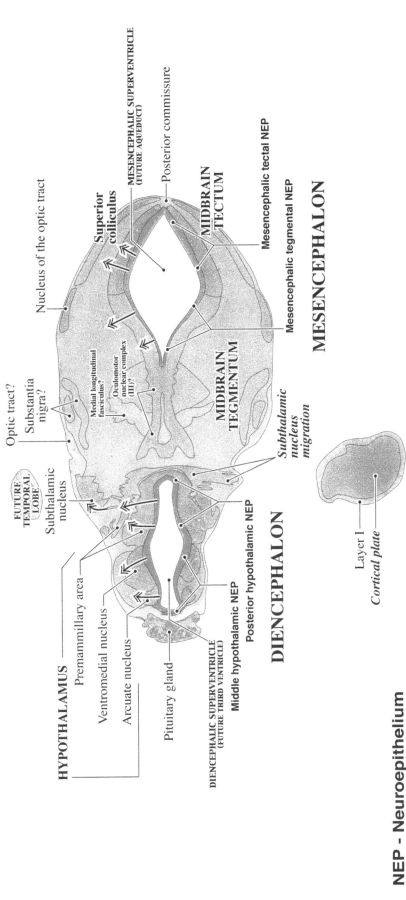

MESENCEPHALIC SUPERVENTRICLE
(FUTURE AQUEDUCT)

Posterior commissure

Nucleus of the optic tract

**MIDBRAIN
TECTUM**

**Superior
colliculus**

Mesencephalic tectal NEP

Mesencephalic tegmental NEP

MESENCEPHALON

Optic tract?

Substantia
nigra?

Medial longitudinal
fasciculus?

Oculomotor
nuclear complex
(III)?

**MIDBRAIN
TEGMENTUM**

*Subthalamic
nucleus
migration*

FUTURE
TEMPORAL
LOBE

Subthalamic
nucleus

HYPOTHALAMUS

Premammillary area

Ventromedial nucleus

Arcuate nucleus

Pituitary gland

DIENCEPHALIC SUPERVENTRICLE
(FUTURE THIRD VENTRICLE)

Middle hypothalamic NEP

Posterior hypothalamic NEP

DIENCEPHALON

Layer I

Cortical plate

BASE OF TELENCEPHALON

NEP – Neuroepithelium

Arrows indicate the
presumed *direction of
neuron migration* from
neuroepithelial sources.

FONT KEY:
VENTRICULAR DIVISIONS – CAPITALS
Germinal zone - Helvetica bold
Transient structure - Times bold italic
Permanent structure - Times Roman or **Bold**

92

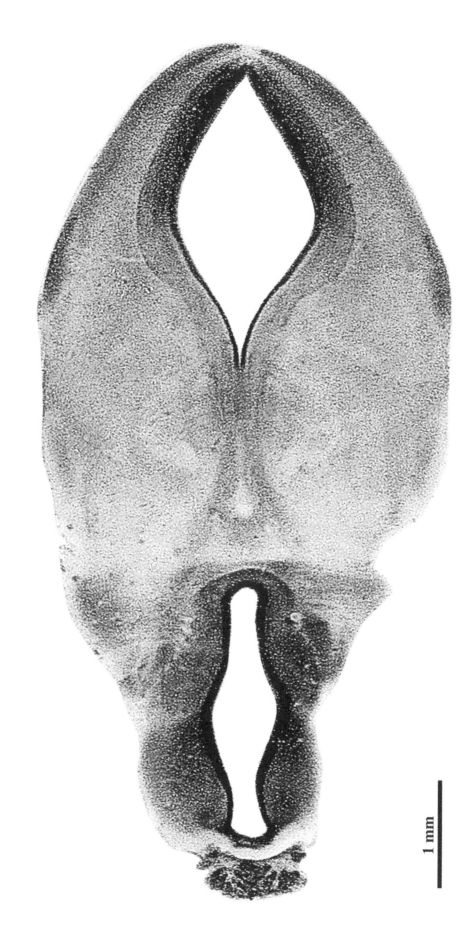

PLATE 30C
CR 43 mm, GW10.6, C886
Horizontal/Frontal
Section 90

1 mm

See the entire section in parts A and B of this plate on the preceding pages.

PLATE 30D

93

Superior colliculus

Successive waves of migrating superior colliculus neurons

Nucleus of the optic tract

Posterior commissure?

MIDBRAIN TECTUM

MESENCEPHALIC SUPERVENTRICLE
(FUTURE CEREBRAL AQUEDUCT)

Mesencephalic tectal NEP

Central gray

Migrating and settling tegmental neurons

Reticular formation

Mesencephalic tegmental NEP

Optic tract?

Substantia nigra?

Medial longitudinal fasciculus?

Incipient cerebral peduncle?

Oculomotor nuclear complex (III)?

MIDBRAIN TEGMENTUM

Reticular formation

Subthalamic nucleus

M E S E N C E P H A L O N

Pituitary gland

Hypothalamic migration

DIENCEPHALIC SUPERVENTRICLE
(FUTURE THIRD VENTRICLE)

Hypothalamic neuroepithelium (NEP)

Arcuate nucleus

Ventromedial nucleus

Premammillary area

HYPOTHALAMUS

D I E N C E P H A L O N

NEP - Neuroepithelium

Arrows indicate the presumed *direction of neuron migration* from neuroepithelial sources.

Arrows indicate the presumed *direction of axon growth* in brain fiber tracts.

FONT KEY:
VENTRICULAR DIVISIONS – CAPITALS
Germinal zone - Helvetica bold
Transient structure - Times bold italic
Permanent structure - Times Roman or **Bold**

94

PLATE 31A
CR 43 mm, GW10.6, C886
Horizontal/Frontal
Section 100

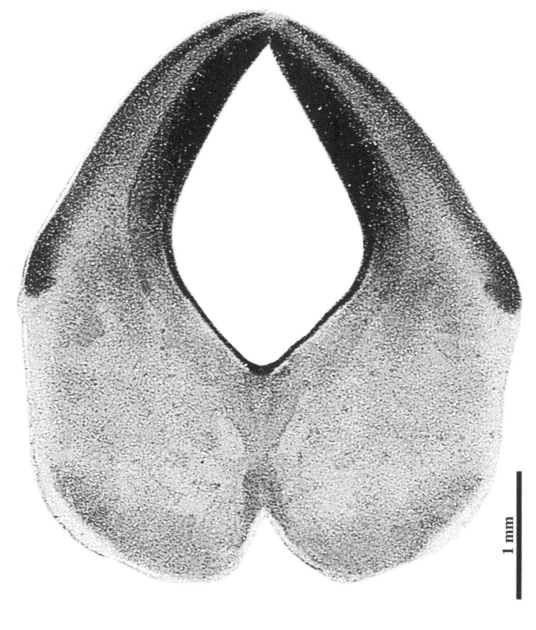

1 mm

Plates 31 to 38 are only shown at higher magnification.

PLATE 31B

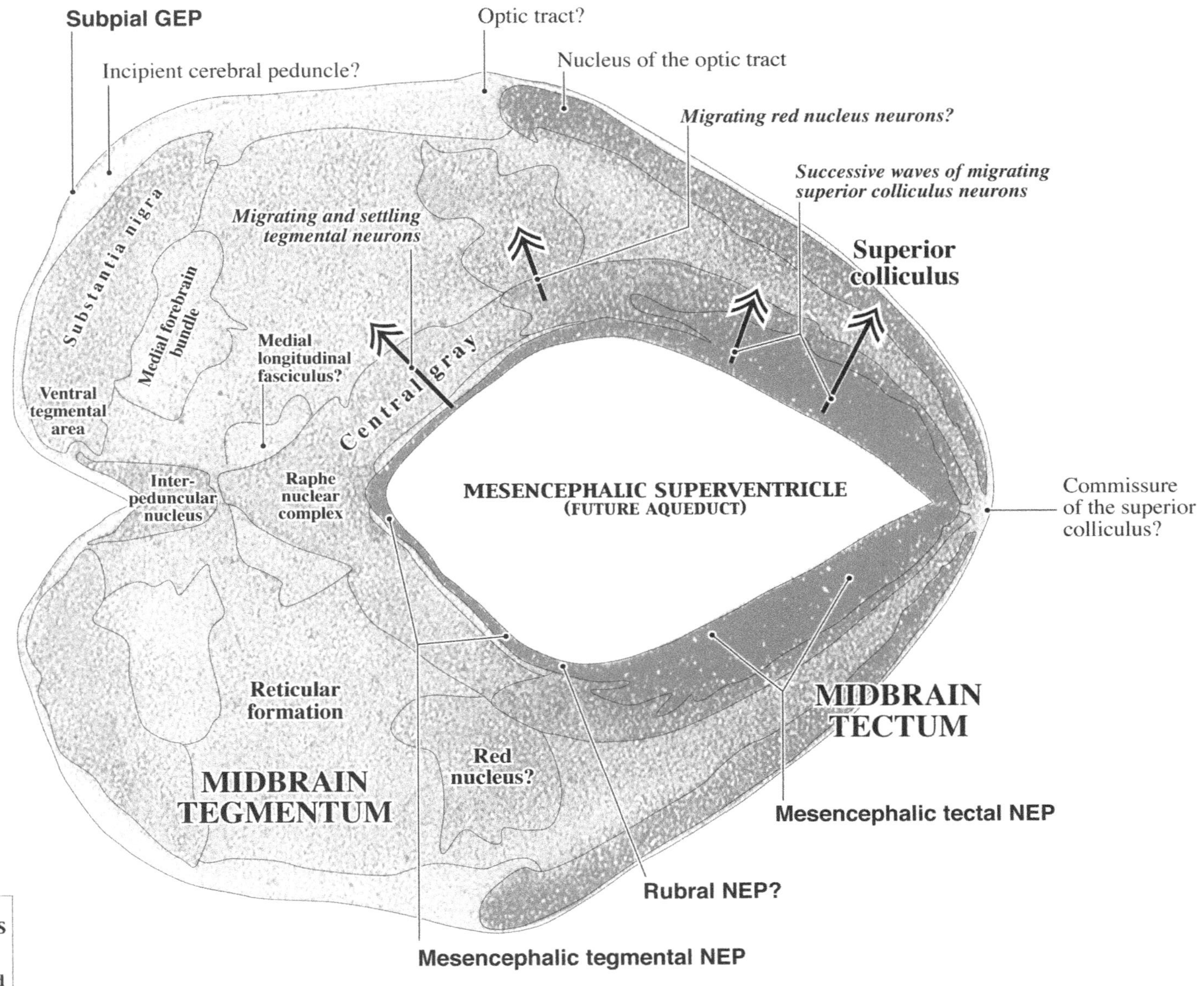

Commissure
of the superior
colliculus?

**Superior
colliculus**

*Successive waves of migrating
superior colliculus neurons*

Migrating red nucleus neurons?

Nucleus of the optic tract

Optic tract?

Incipient cerebral peduncle?

Subpial GEP

Substantia nigra

Ventral
tegmental
area

*Medial forebrain
bundle*

Inter-
peduncular
nucleus

*Migrating and settling
tegmental neurons*

Medial
longitudinal
fasciculus?

Raphe
nuclear
complex

Central gray

MESENCEPHALIC SUPERVENTRICLE
(FUTURE AQUEDUCT)

**MIDBRAIN
TECTUM**

Mesencephalic tectal NEP

Rubral NEP?

Red
nucleus?

Mesencephalic tegmental NEP

Reticular
formation

**MIDBRAIN
TEGMENTUM**

ABBREVIATIONS:
GEP - Glioepithelium
NEP - Neuroepithelium

Arrows indicate the
presumed *direction of*
neuron migration from
neuroepithelial sources.

FONT KEY:
VENTRICULAR DIVISIONS - CAPITALS
Germinal zone - Helvetica bold
Transient structure - Times bold italic
Permanent structure - Times Roman or **Bold**

96

PLATE 32A
CR 43 mm, GW10.6, C886
Horizontal/Frontal
Section 105

Plates 31 to 38 are only shown at higher magnification.

1 mm

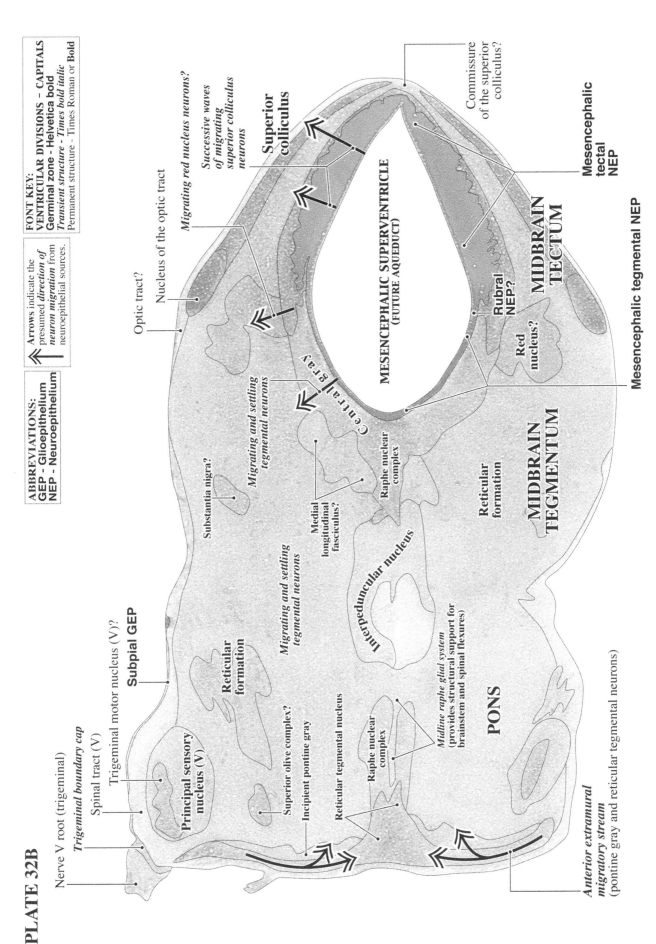

PLATE 32B

Nerve V root (trigeminal)

Trigeminal boundary cap

Spinal tract (V)

Trigeminal motor nucleus (V)?

Subpial GEP

**Principal sensory
nucleus (V)**

**Reticular
formation**

Superior olive complex?

Incipient pontine gray

Reticular tegmental nucleus

Raphe nuclear
complex

*Midline raphe glial system
(provides structural support for
brainstem and spinal flexures)*

PONS

*Anterior extramural
migratory stream*
(pontine gray and reticular tegmental neurons)

Substantia nigra?

*Migrating and settling
tegmental neurons*

*Migrating and settling
tegmental neurons*

Medial
longitudinal
fasciculus?

Interpeduncular nucleus

Raphe nuclear
complex

Central gray

Reticular
formation

**MIDBRAIN
TEGMENTUM**

Red
nucleus?

**Rubral
NEP?**

**MIDBRAIN
TECTUM**

Optic tract?

Nucleus of the optic tract

Migrating red nucleus neurons?

*Successive waves
of migrating
superior colliculus
neurons*

**Superior
colliculus**

MESENCEPHALIC SUPERVENTRICLE
(FUTURE AQUEDUCT)

Commissure
of the superior
colliculus?

**Mesencephalic
tectal
NEP**

Mesencephalic tegmental NEP

ABBREVIATIONS:
GEP - Glioepithelium
NEP - Neuroepithelium

FONT KEY:
VENTRICULAR DIVISIONS – CAPITALS
Germinal zone - Helvetica bold
Transient structure - Times bold italic
Permanent structure - Times Roman or **Bold**

Arrows indicate the
presumed *direction of
neuron migration* from
neuroepithelial sources.

98

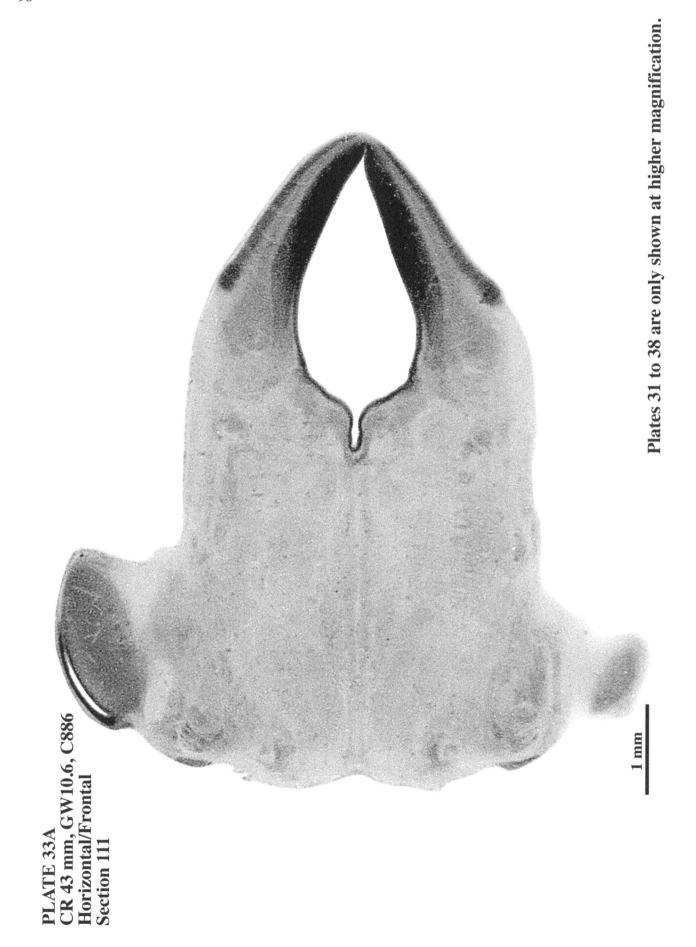

PLATE 33A
CR 43 mm, GW10.6, C886
Horizontal/Frontal
Section 111

Plates 31 to 38 are only shown at higher magnification.

1 mm

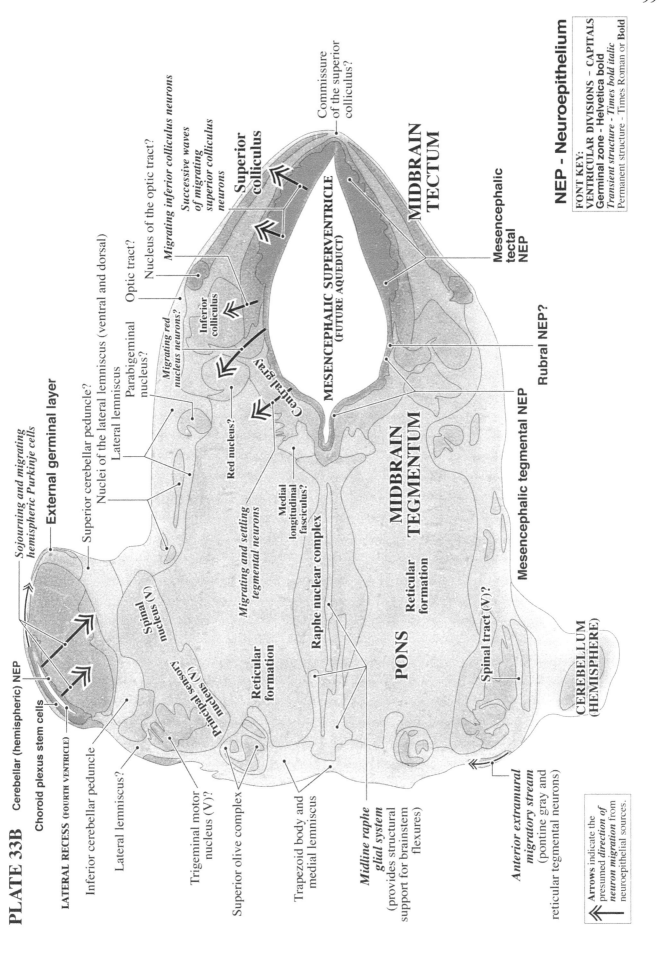

PLATE 33B Cerebellar (hemispheric) NEP

Sojourning and migrating hemispheric Purkinje cells

Choroid plexus stem cells

External germinal layer

LATERAL RECESS (FOURTH VENTRICLE)

LATERAL (hemispheric) NEP

Inferior cerebellar peduncle

Lateral lemniscus

Superior cerebellar peduncle?
Nuclei of the lateral lemniscus (ventral and dorsal)

Lateral lemniscus

Parabigeminal nucleus?

Migrating red nucleus neurons?

Inferior colliculus

Optic tract?

Nucleus of the optic tract?

Migrating inferior colliculus neurons

Successive waves of migrating superior colliculus neurons

Superior colliculus

Commissure of the superior colliculus?

MIDBRAIN TECTUM

MESENCEPHALIC SUPERVENTRICLE
(FUTURE AQUEDUCT)

Mesencephalic tectal NEP

Rubral NEP?

Mesencephalic tegmental NEP

MIDBRAIN TEGMENTUM

Red nucleus?

Central gray

Spinal nucleus (V)

Trigeminal motor nucleus (V)?

Principal sensory nucleus (V)?

Superior olive complex

Reticular formation

Migrating and settling tegmental neurons

Medial longitudinal fasciculus?

Raphe nuclear complex

MIDBRAIN TEGMENTUM

PONS

Reticular formation

Trapezoid body and medial lemniscus

Midline raphe glial system
(provides structural support for brainstem flexures)

Spinal tract (V)?

CEREBELLUM (HEMISPHERE)

Anterior extramural migratory stream
(pontine gray and reticular tegmental neurons)

NEP - Neuroepithelium

FONT KEY:
VENTRICULAR DIVISIONS – CAPITALS
Germinal zone - Helvetica bold
Transient structure - Times bold italic
Permanent structure - Times Roman or **Bold**

Arrows indicate the presumed *direction of neuron migration* from neuroepithelial sources.

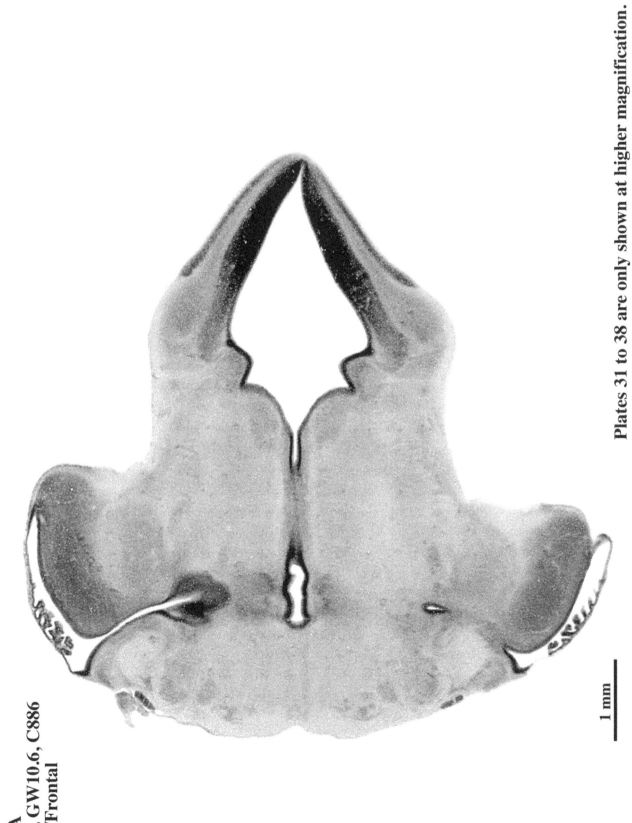

PLATE 34A
CR 43 mm, GW10.6, C886
Horizontal/Frontal
Section 117

Plates 31 to 38 are only shown at higher magnification.

1 mm

PLATE 34B

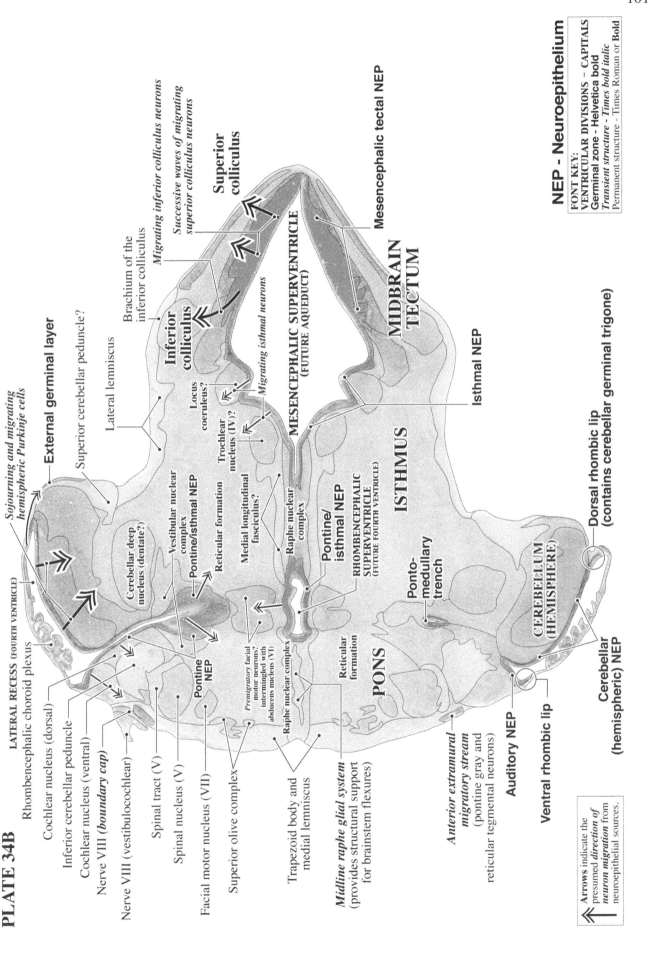

101

Rhombencephalic choroid plexus

LATERAL RECESS (FOURTH VENTRICLE)

Sojourning and migrating hemispheric Purkinje cells

External germinal layer

Superior cerebellar peduncle?

Migrating inferior colliculus neurons

Successive waves of migrating superior colliculus neurons

Superior colliculus

Brachium of the inferior colliculus

Lateral lemniscus

Mesencephalic tectal NEP

Inferior colliculus

Locus coeruleus?

Migrating isthmal neurons

MESENCEPHALIC SUPERVENTRICLE
(FUTURE AQUEDUCT)

MIDBRAIN TECTUM

Cochlear nucleus (dorsal)

Inferior cerebellar peduncle

Cochlear nucleus (ventral)

Nerve VIII (*boundary cap*)

Nerve VIII (vestibulocochlear)

Cerebellar deep nucleus (dentate?)

Vestibular nuclear complex

Pontine/isthmal NEP

Reticular formation

Trochlear nucleus (IV)?

Medial longitudinal fasciculus?

Raphe nuclear complex

Pontine/isthmal NEP

RHOMBENCEPHALIC SUPERVENTRICLE
(FUTURE FOURTH VENTRICLE)

Isthmal NEP

ISTHMUS

Spinal tract (V)

Spinal nucleus (V)

Facial motor nucleus (VII)

Superior olive complex

Pontine NEP

Premigratory facial motor neurons? intermingled with abducens nucleus (VI)

Raphe nuclear complex

Reticular formation

Ponto-medullary trench

PONS

Trapezoid body and medial lemniscus

Midline raphe glial system
(provides structural support for brainstem flexures)

CEREBELLUM (HEMISPHERE)

Auditory NEP

Anterior extramural migratory stream
(pontine gray and reticular tegmental neurons)

Ventral rhombic lip

Cerebellar (hemispheric) NEP

Dorsal rhombic lip
(contains cerebellar germinal trigone)

NEP - Neuroepithelium

FONT KEY:
VENTRICULAR DIVISIONS – CAPITALS
Germinal zone - Helvetica bold
Transient structure - Times bold italic
Permanent structure - Times Roman or **Bold**

Arrows indicate the presumed *direction of neuron migration* from neuroepithelial sources.

102

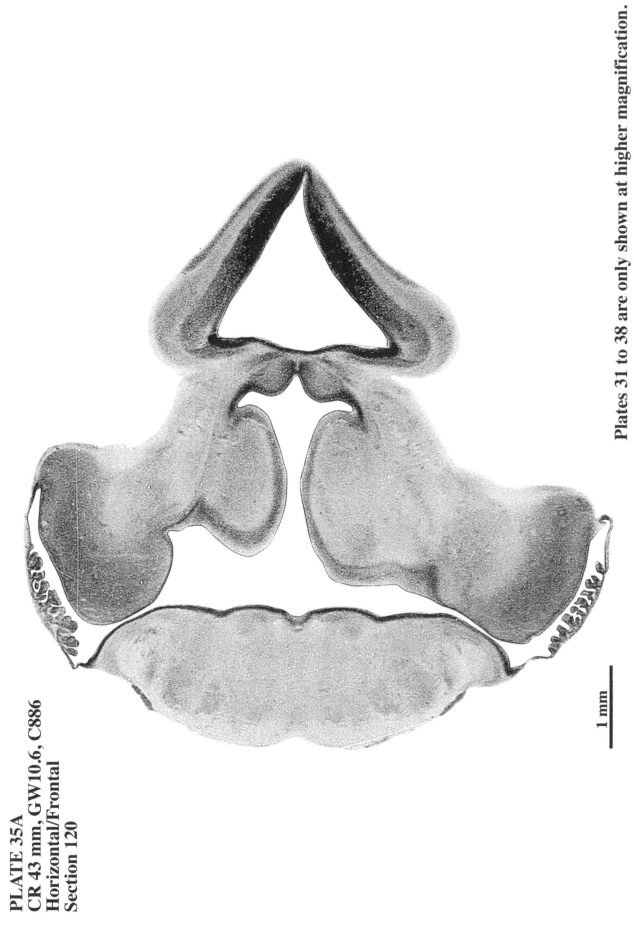

PLATE 35A
CR 43 mm, GW10.6, C886
Horizontal/Frontal
Section 120

Plates 31 to 38 are only shown at higher magnification.

1 mm

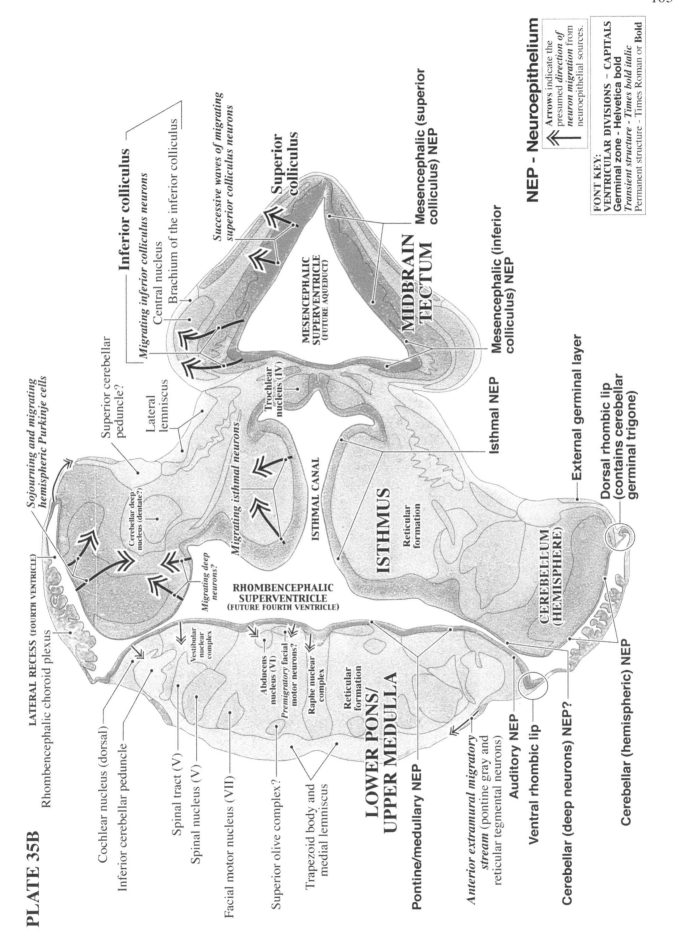

PLATE 35B

LATERAL RECESS (FOURTH VENTRICLE)
Rhombencephalic choroid plexus

Inferior colliculus

Successive waves of migrating superior colliculus neurons

Migrating inferior colliculus neurons

Central nucleus

Brachium of the inferior colliculus

Superior colliculus

Mesencephalic (superior colliculus) NEP

MESENCEPHALIC SUPERVENTRICLE (FUTURE AQUEDUCT)

MIDBRAIN TECTUM

Mesencephalic (inferior colliculus) NEP

Sojourning and migrating hemispheric Purkinje cells

Superior cerebellar peduncle?

Lateral lemniscus

Trochlear nucleus (IV)

Isthmal NEP

Cerebellar deep nuclei (dentate?)

Migrating isthmal neurons

ISTHMAL CANAL

ISTHMUS

Reticular formation

External germinal layer

Dorsal rhombic lip (contains cerebellar germinal trigone)

RHOMBENCEPHALIC SUPERVENTRICLE (FUTURE FOURTH VENTRICLE)

Migrating deep neurons?

CEREBELLUM (HEMISPHERE)

Cochlear nucleus (dorsal)

Inferior cerebellar peduncle

Spinal tract (V)

Spinal nucleus (V)

Facial motor nucleus (VII)

Vestibular nuclear complex

Abducens nucleus (VI)

Premigratory facial motor neurons?

Raphe nuclear complex

Reticular formation

LOWER PONS/ UPPER MEDULLA

Pontine/medullary NEP

Superior olive complex?

Trapezoid body and medial lemniscus

Anterior extramural migratory stream (pontine gray and reticular tegmental neurons)

Auditory NEP

Ventral rhombic lip

Cerebellar (deep neurons) NEP?

Cerebellar (hemispheric) NEP

NEP - Neuroepithelium

Arrows indicate the presumed direction of neuron migration from neuroepithelial sources.

FONT KEY:
VENTRICULAR DIVISIONS – CAPITALS
Germinal zone - Helvetica bold
Transient structure - Times bold italic
Permanent structure - Times Roman or Bold

103

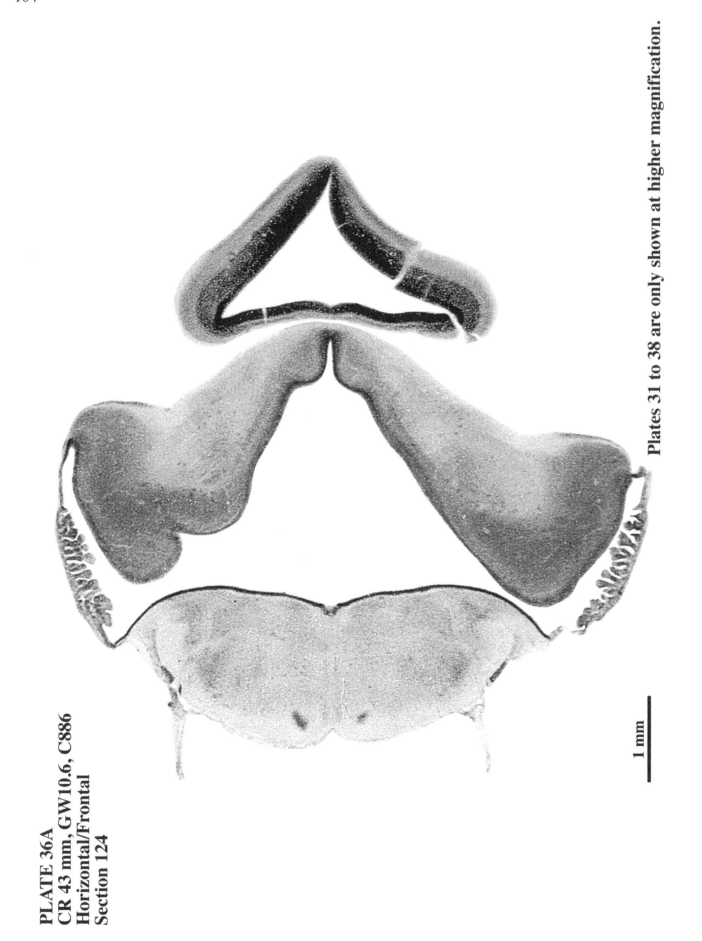

PLATE 36A
CR 43 mm, GW10.6, C886
Horizontal/Frontal
Section 124

Plates 31 to 38 are only shown at higher magnification.

1 mm

PLATE 36B

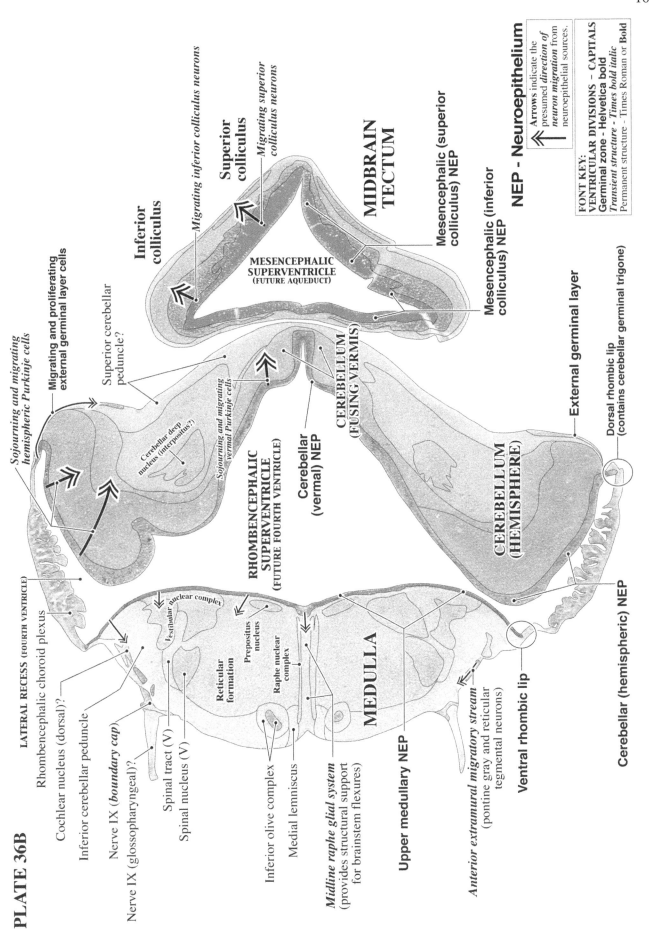

LATERAL RECESS (FOURTH VENTRICLE)

Rhombencephalic choroid plexus

Cochlear nucleus (dorsal)?

Inferior cerebellar peduncle

Nerve IX (*boundary cap*)?

Nerve IX (glossopharyngeal)?

Spinal tract (V)

Spinal nucleus (V)

Inferior olive complex

Medial lemniscus

Midline raphe glial system
(provides structural support
for brainstem flexures)

Upper medullary NEP

Anterior extramural migratory stream
(pontine gray and reticular
tegmental neurons)

Cerebellar (hemispheric) NEP

*Sojourning and migrating
hemispheric Purkinje cells*

**Migrating and proliferating
external germinal layer cells**

Superior cerebellar
peduncle?

Cerebellar deep
nucleus (interpositus?)

*Sojourning and migrating
vermal Purkinje cells*

**CEREBELLUM
(FUSING VERMIS)**

**Cerebellar
(vermal) NEP**

**RHOMBENCEPHALIC
SUPERVENTRICLE**
(FUTURE FOURTH VENTRICLE)

**CEREBELLUM
(HEMISPHERE)**

External germinal layer

Dorsal rhombic lip
(contains cerebellar germinal trigone)

Migrating inferior colliculus neurons

*Migrating superior
colliculus neurons*

**Superior
colliculus**

**Inferior
colliculus**

**MESENCEPHALIC
SUPERVENTRICLE**
(FUTURE AQUEDUCT)

**MIDBRAIN
TECTUM**

**Mesencephalic (superior
colliculus) NEP**

**Mesencephalic (inferior
colliculus) NEP**

Vestibular nuclear complex

Prepositus
nucleus

**Reticular
formation**

**Raphe nuclear
complex**

MEDULLA

Ventral rhombic lip

NEP - Neuroepithelium

Arrows indicate the
presumed *direction of
neuron migration* from
neuroepithelial sources.

FONT KEY:
VENTRICULAR DIVISIONS – CAPITALS
Germinal zone - Helvetica bold
Transient structure - Times bold italic
Permanent structure - Times Roman or **Bold**

106

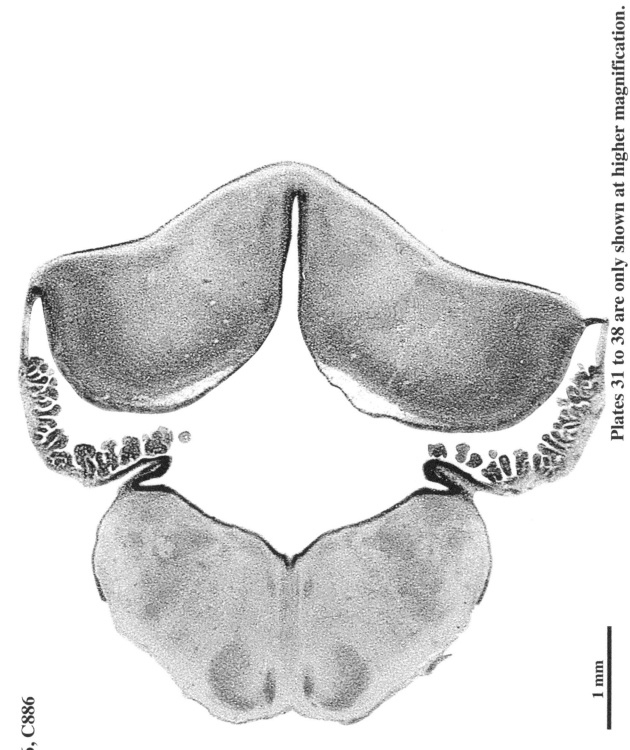

PLATE 37A
CR 43 mm, GW10.6, C886
Horizontal/Frontal
Section 134

Plates 31 to 38 are only shown at higher magnification.

1 mm

PLATE 37B

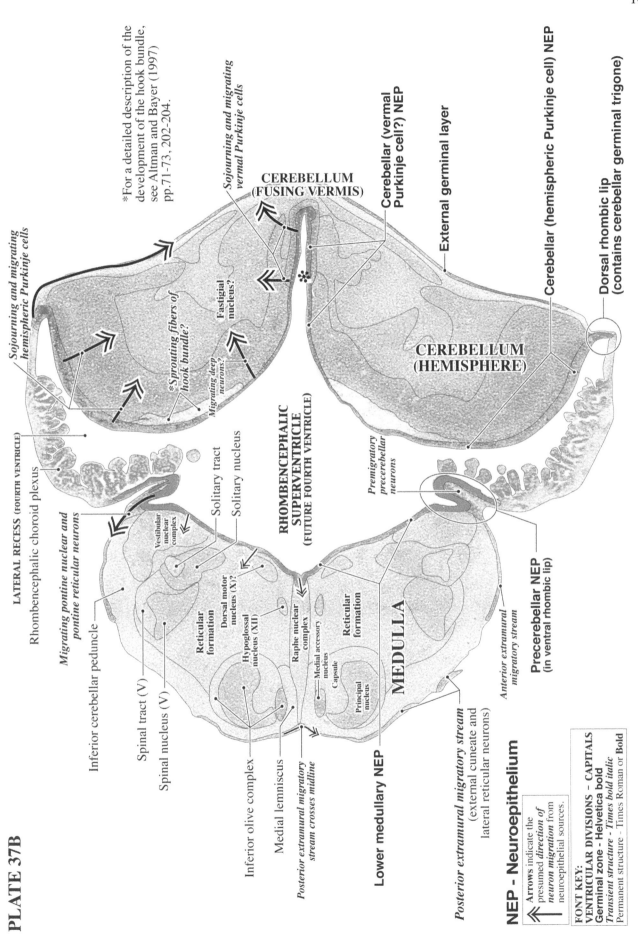

*For a detailed description of the development of the hook bundle, see Altman and Bayer (1997) pp.71-73, 202-204.

Sojourning and migrating hemispheric Purkinje cells

Sojourning and migrating vermal Purkinje cells

CEREBELLUM (FUSING VERMIS)

Fastigial nucleus?

Cerebellar (vermal Purkinje cell?) NEP

External germinal layer

Cerebellar (hemispheric Purkinje cell) NEP

Dorsal rhombic lip (contains cerebellar germinal trigone)

*Sprouting fibers of hook bundle?

Migrating deep neurons?

CEREBELLUM (HEMISPHERE)

Premigratory precerebellar neurons

LATERAL RECESS (FOURTH VENTRICLE)
Rhombencephalic choroid plexus

Migrating pontine nuclear and pontine reticular neurons

Vestibular nuclear complex

Solitary tract
Solitary nucleus

RHOMBENCEPHALIC SUPERVENTRICLE
(FUTURE FOURTH VENTRICLE)

Inferior cerebellar peduncle

Reticular formation

Dorsal motor nucleus (X)?

Hypoglossal nucleus (XII)

Raphe nuclear complex

Medial accessory nucleus

Capsule

Principal nucleus

Reticular formation

MEDULLA

Spinal tract (V)
Spinal nucleus (V)

Inferior olive complex

Medial lemniscus

Posterior extramural migratory stream crosses midline

Anterior extramural migratory stream

Precerebellar NEP
(in ventral rhombic lip)

Posterior extramural migratory stream (external cuneate and lateral reticular neurons)

Lower medullary NEP

NEP - Neuroepithelium

Arrows indicate the presumed *direction of neuron migration* from neuroepithelial sources.

FONT KEY:
VENTRICULAR DIVISIONS - CAPITALS
Germinal zone - Helvetica bold
Transient structure - Times bold italic
Permanent structure - Times Roman or **Bold**

108

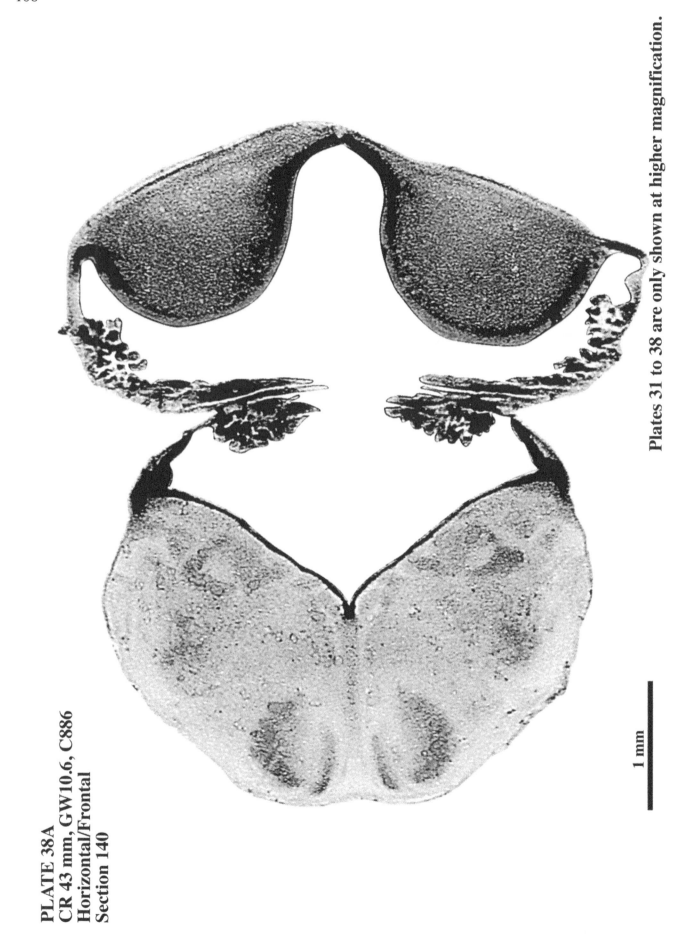

PLATE 38A
CR 43 mm, GW10.6, C886
Horizontal/Frontal
Section 140

Plates 31 to 38 are only shown at higher magnification.

1 mm

109

PLATE 38B

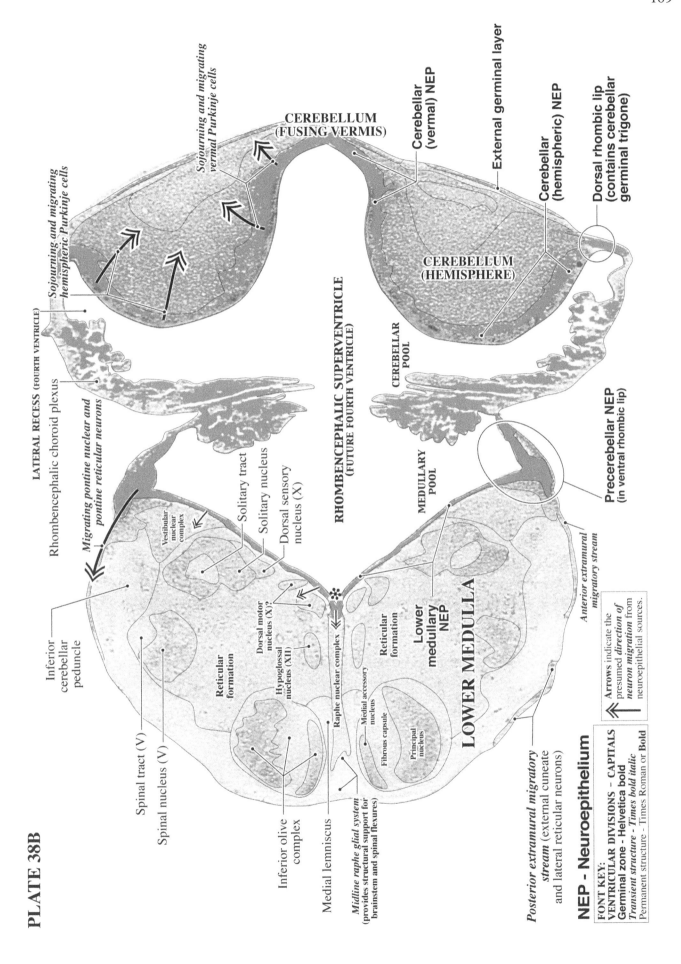

Sojourning and migrating vermal Purkinje cells

Sojourning and migrating hemispheric Purkinje cells

CEREBELLUM (FUSING VERMIS)

Cerebellar (vermal) NEP

External germinal layer

Cerebellar (hemispheric) NEP

Dorsal rhombic lip (contains cerebellar germinal trigone)

CEREBELLUM (HEMISPHERE)

CEREBELLAR POOL

LATERAL RECESS (FOURTH VENTRICLE)

Rhombencephalic choroid plexus

RHOMBENCEPHALIC SUPERVENTRICLE (FUTURE FOURTH VENTRICLE)

MEDULLARY POOL

Precerebellar NEP (in ventral rhombic lip)

Migrating pontine nuclear and pontine reticular neurons

Vestibular nuclear complex

Solitary tract

Solitary nucleus

Dorsal sensory nucleus (X)

Inferior cerebellar peduncle

Spinal tract (V)

Spinal nucleus (V)

Reticular formation

Dorsal motor nucleus (X)?

Hypoglossal nucleus (XII)

Raphe nuclear complex

Medial accessory nucleus

Fibrous capsule

Principal nucleus

Reticular formation

Lower medullary NEP

LOWER MEDULLA

Inferior olive complex

Medial lemniscus

Midline raphe glial system (provides structural support for brainstem and spinal flexures)

Posterior extramural migratory stream (external cuneate and lateral reticular neurons)

Anterior extramural migratory stream

Arrows indicate the presumed direction of neuron migration from neuroepithelial sources.

NEP - Neuroepithelium

FONT KEY:
VENTRICULAR DIVISIONS - CAPITALS
Germinal zone - Helvetica bold
Transient structure - Times bold italic
Permanent structure - Times Roman or Bold

PART IV: C6658
CR 40 mm (GW 10.3)
Sagittal

This is specimen 6658 in the Carnegie Collection, designated here as C6658, a female with a crown-rump length (CR) of 40 mm estimated to be at gestational week (GW) 10.3. The entire fetus was cut in the sagittal plane in 40-μm sections and stained with hematoxylin and eosin. Information on the date of specimen collection, fixative, and embedding medium (appears to be celloidin) was not available to us. The histology is excellent, and this is one of the best preserved specimens in any of the collections at the National Museum of Health and Medicine. Since there is no photograph of this specimen's brain before histological processing, a specimen from Hochstetter (1919) that is comparable in age to C6658 is used to show external brain features at GW 10.3 (**A, Fig. 11**). C6658's brain structures are more difficult to understand because the sections are not cut parallel to the midline; **Figure 11** shows the approximate rotations in horizontal (**B**) and vertical (**C**) dimensions. Photographs of 11 sections are illustrated at low magnification in four parts (**Plates 39A–D** through **49A–D**). The **A** and **B** parts show the brain in place in the skull; the **C** and **D** parts show only the brain (and some peripheral ganglia) at slightly higher magnification. **Plates 50–65** show high-magnification views of various parts of the brain at different levels from the cerebral cortex. Most of the high-magnification plates are rotated 90° (landscape orientation) to more efficiently use page space.

Throughout the cerebral cortex, the neuroepithelium is prominent and appears to be without a subventricular zone, but one cannot tell because of the section thickness. The stratified transitional field (STF) contains STF1 and STF5 throughout, with STF4 only in lateral areas. The most prominent developmental feature of the cerebral cortex is that both the STF layers and the cortical plate have a pronounced anterolateral (thicker) to dorsomedial (thinner) maturation gradient. In anterolateral parts of the cerebral cortex, streams of neurons and glia appear to leave STF4 and enter the lateral migratory stream. The hippocampus contains ammonic and dentate migrations, but there is no evidence of a pyramidal layer in Ammon's horn or

a granule cell layer in the dentate gyrus. A massive neuroepithelium/subventricular zone overlies the amygdala, nucleus accumbens, and striatum (caudate and putamen) where neurons (and glia) are being generated. The olfactory bulb is evaginating in front of the basal telencephalic neuroepithelium and mitral cells are migrating towards it from their origin in the basal telencephalon. The neuroepithelium is still generating the latest neurons in the preoptic area, hypothalamus, and thalamus that will settle near the ventricles.

The midbrain tegmentum, pons, and medulla have the thinnest neuroepithelia that are being transformed into the ependymal lining of the ventricles. The thick precerebellar neuroepithelium is an exception in the medulla where pontine nuclear neurons are being generated. Thicker neuroepithelia are in the midbrain tectum, where many neuronal populations are having their last period of neurogenesis.

Neurons throughout the entire brain are migrating and settling. Many cells have settled in the basal telencephalon and nuclear arrays can be identified. Nuclear divisions are indistinct throughout the diencephalon because massive migrations blur borders. More definition is seen in the midbrain tegmentum, pons, and medulla. The large anterior extramural and posterior extramural migratory streams are prominent in the medulla and pons.

The cerebellum is a thick, smooth plate overlying the posterior pons and medulla, and a definite neuroepithelium at the ventricular surface (the progenitors of Golgi cells are stockbuilding). Many Purkinje cells are sojourning in a dense layer outside the neuroepithelium, and others are migrating upward. Many of the deep neurons are superficial in the cerebellum, but some are migrating downward to intermingle with upwardly migrating Purkinje cells. The cortical surface is partially covered by an external germinal layer (egl) that is actively producing neuronal stem cells, as it grows over the surface of the cerebellar cortex.

GW 10.3 SAGITTAL

A perfect sagittal cut through the brain bisects the cerebral cortex into two separate hemispheres by passing through the interhemispheric fissure, and does the same in the brainstem by passing through the midline of the ventricles.

Sections of C6658's brain are not parallel to the midline either horizontally (-11.71°, top view) or vertically (-6.64°, back view). In each of the illustrated sections on the following pages, the anterior edge of the cortex (top right) is tilted away from the observer, while the medulla and upper spinal cord (bottom) are tilted toward the observer.

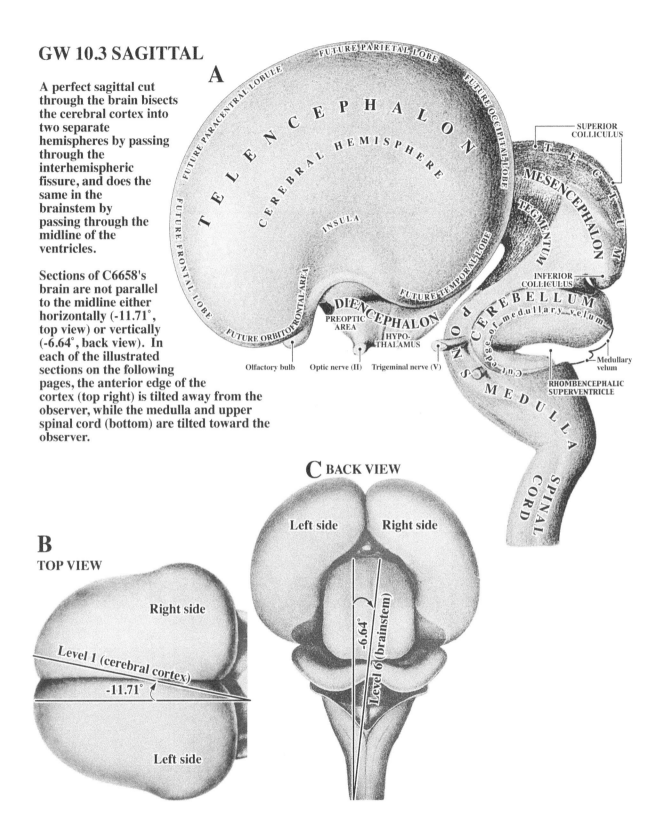

A

FUTURE PARACENTRAL LOBULE

FUTURE PARIETAL LOBE

FUTURE OCCIPITAL LOBE

T E L E N C E P H A L O N

CEREBRAL HEMISPHERE

MESENCEPHALON

SUPERIOR COLLICULUS

T E C T U M

TEGMENTUM

INSULA

FUTURE FRONTAL LOBE

INFERIOR COLLICULUS

FUTURE TEMPORAL LOBE

CEREBELLUM

Cut edge of medullary velum

DIENCEPHALON

PREOPTIC AREA

HYPO-THALAMUS

P O N S

Medullary velum

FUTURE ORBITOFRONTAL AREA

MEDULLA

RHOMBENCEPHALIC SUPERVENTRICLE

Olfactory bulb   Optic nerve (II)   Trigeminal nerve (V)

SPINAL CORD

C BACK VIEW

Left side   Right side

-6.64°

Level 6 (brainstem)

B

TOP VIEW

Right side

Level 1 (cerebral cortex)

-11.71°

Left side

Figure 11. A, Lateral view of the brain and upper cervical spinal cord from a specimen with a crown-rump length of 38 mm (modified from Figure 43, Table VII, Hochstetter, 1919) identifies external features of a brain similar to C6658 (CR 40 mm). **B,** Top view of the brain in **A** (modified from Figure 45, Table VIII, Hochstetter, 1919) shows how C6658's sections rotate from a line parallel to the horizontal midline in the interhemispheric fissure. **C,** Back view of the brain in **A** (modified from Figure 44, Table VIII, Hochstetter, 1919) shows how C6658's sections rotate from a line parallel to the vertical midline in the brainstem and upper cervical spinal cord.

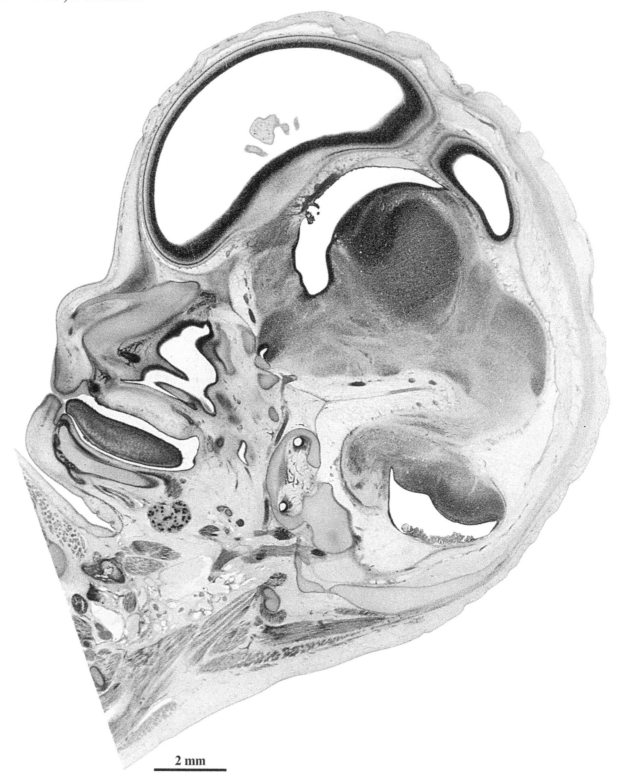

2 mm

Neuroepithelial divisions, glioepithelial divisions, and differentiating
structures are labeled in Parts C and D of this plate on the following pages.

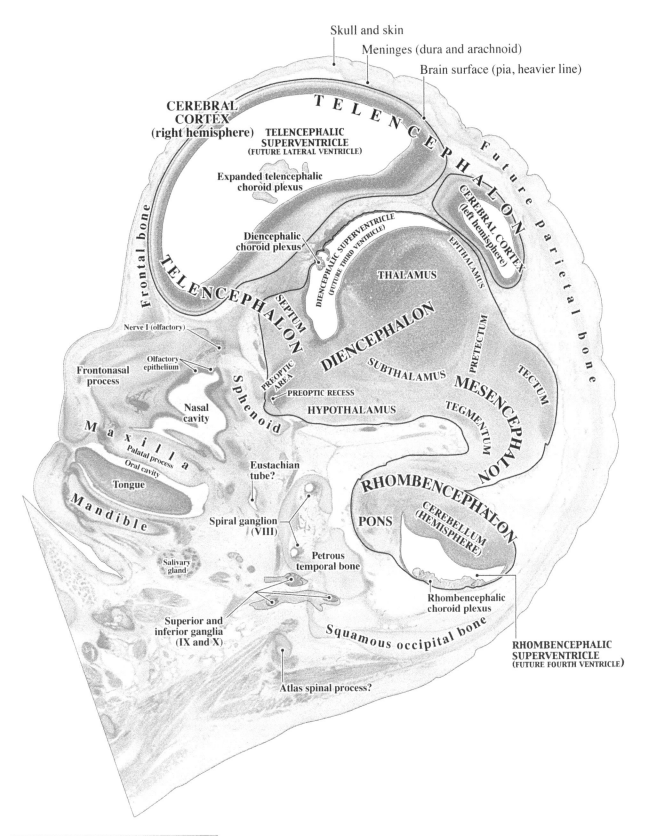

Skull and skin

Meninges (dura and arachnoid)

Brain surface (pia, heavier line)

CEREBRAL CORTEX
(right hemisphere)

TELENCEPHALIC SUPERVENTRICLE
(FUTURE LATERAL VENTRICLE)

T E L E N C E P H A L O N

F u t u r e   p a r i e t a l   b o n e

Expanded telencephalic choroid plexus

CEREBRAL CORTEX
(left hemisphere)

Diencephalic choroid plexus

DIENCEPHALIC SUPERVENTRICLE
(FUTURE THIRD VENTRICLE)

EPITHALAMUS

THALAMUS

Frontal bone

TELENCEPHALON

Nerve I (olfactory)

Olfactory epithelium

Frontonasal process

SEPTUM

PREOPTIC AREA

PRETECTUM

DIENCEPHALON

SUBTHALAMUS

PREOPTIC RECESS

TECTUM

MESENCEPHALON

HYPOTHALAMUS

TEGMENTUM

Sphenoid

Nasal cavity

M a x i l l a

Palatal process

Oral cavity

Tongue

Eustachian tube?

RHOMBENCEPHALON

M a n d i b l e

Spiral ganglion (VIII)

PONS

CEREBELLUM
(HEMISPHERE)

Salivary gland

Petrous temporal bone

Superior and inferior ganglia
(IX and X)

Rhombencephalic choroid plexus

S q u a m o u s   o c c i p i t a l   b o n e

RHOMBENCEPHALIC SUPERVENTRICLE
(FUTURE FOURTH VENTRICLE)

Atlas spinal process?

FONT KEY:
VENTRICULAR DIVISIONS – CAPITALS
Major brain structure - Times **Bold CAPITALS**
All other structures - Times Roman or **Bold**

PLATE 39C
CR 40 mm, GW 10.3, C6658
Sagittal
Slide 53, Section 1

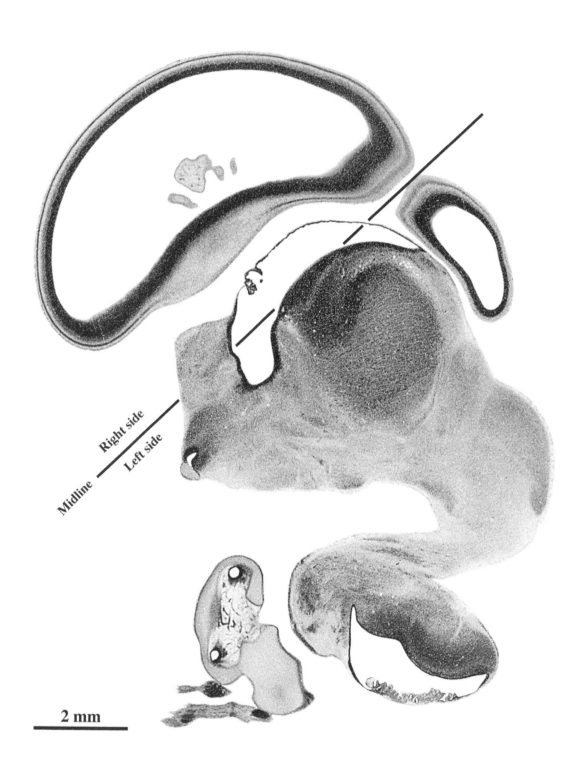

Right side

Left side

Midline

2 mm

The head, major brain structures, and ventricular divisions are
labeled in Parts A and B of this plate on the preceding pages.

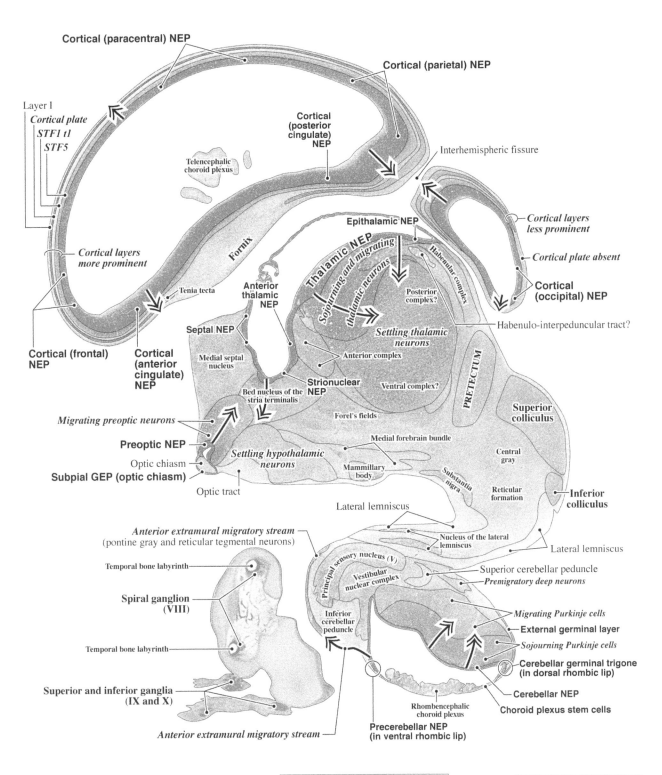

Cortical (paracentral) NEP

Cortical (parietal) NEP

Layer I
Cortical plate
STF1 t1
STF5

Cortical
(posterior
cingulate)
NEP

*Telencephalic
choroid plexus*

Interhemispheric fissure

Epithalamic NEP

Thalamic NEP

*Sojourning and migrating
thalamic neurons*

*Cortical layers
less prominent*

*Habenular
complex*

Cortical plate absent

Cortical
(occipital) NEP

*Cortical layers
more prominent*

Fornix

Tenia tecta

Anterior
thalamic
NEP

Posterior
complex?

Habenulo-interpeduncular tract?

*Settling thalamic
neurons*

Septal NEP

*Medial septal
nucleus*

Anterior complex

Cortical (frontal)
NEP

Cortical
(anterior
cingulate)
NEP

Strionuclear
NEP

Ventral complex?

PRETECTUM

Superior
colliculus

*Bed nucleus of the
stria terminalis*

Forel's fields

Migrating preoptic neurons

Preoptic NEP

Medial forebrain bundle

Central
gray

*Settling hypothalamic
neurons*

Optic chiasm

Subpial GEP (optic chiasm)

Optic tract

*Mammillary
body*

*Substantia
nigra*

Reticular
formation

Inferior
colliculus

Lateral lemniscus

Anterior extramural migratory stream
(pontine gray and reticular tegmental neurons)

Nucleus of the lateral
lemniscus

Lateral lemniscus

Temporal bone labyrinth

Principal sensory nucleus (V)

*Vestibular
nuclear complex*

Superior cerebellar peduncle

Premigratory deep neurons

Spiral ganglion
(VIII)

Inferior
cerebellar
peduncle

Migrating Purkinje cells

External germinal layer

Sojourning Purkinje cells

Temporal bone labyrinth

Cerebellar germinal trigone
(in dorsal rhombic lip)

Superior and inferior ganglia
(IX and X)

*Rhombencephalic
choroid plexus*

Cerebellar NEP

Choroid plexus stem cells

Anterior extramural migratory stream

Precerebellar NEP
(in ventral rhombic lip)

FONT KEY:
Germinal zone - Helvetica bold
Transient structure - Times bold italic
Permanent structure - Times Roman or **Bold**

ABBREVIATIONS:
GEP - Glioepithelium
NEP - Neuroepithelium
STF - Stratified transitional field

Arrows indicate the
presumed *direction of
neuron migration* from
neuroepithelial sources.

**PLATE 40A
CR 40 mm, GW 10.3, C6658
Sagittal
Slide 63, Section 1**

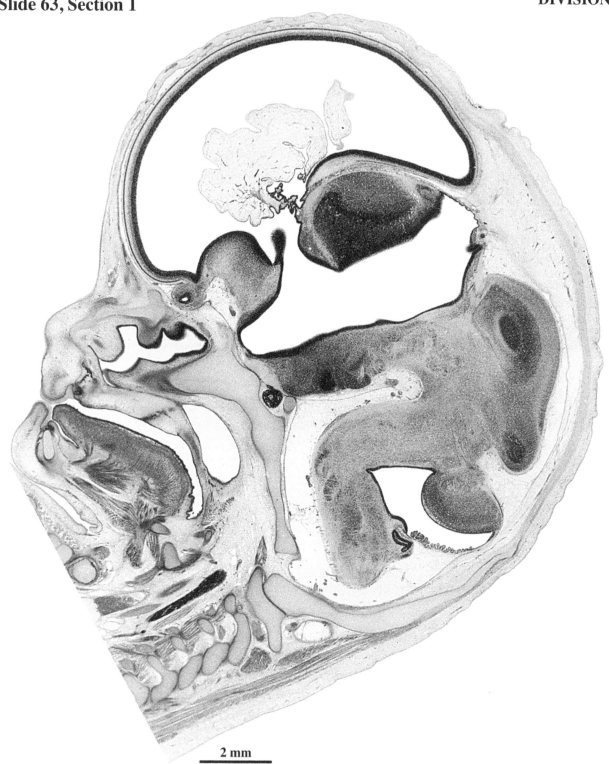

2 mm

Neuroepithelial divisions, glioepithelial divisions, and differentiating
structures are labeled in Parts C and D of this plate on the following pages.

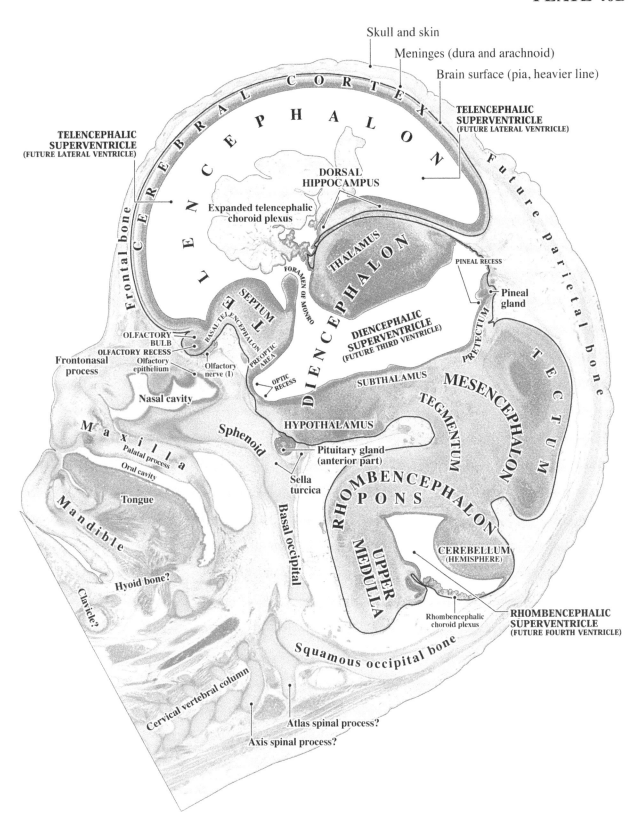

Skull and skin

Meninges (dura and arachnoid)

Brain surface (pia, heavier line)

TELENCEPHALIC SUPERVENTRICLE (FUTURE LATERAL VENTRICLE)

TELENCEPHALIC SUPERVENTRICLE (FUTURE LATERAL VENTRICLE)

CEREBRAL CORTEX

TELENCEPHALON

Future parietal bone

DORSAL HIPPOCAMPUS

Frontal bone

Expanded telencephalic choroid plexus

THALAMUS

DIENCEPHALON

PINEAL RECESS

Pineal gland

FORAMEN OF MONRO

SEPTUM

BASAL TELENCEPHALON

DIENCEPHALIC SUPERVENTRICLE (FUTURE THIRD VENTRICLE)

PRETECTUM

OLFACTORY BULB

OLFACTORY RECESS

Olfactory epithelium

PREOPTIC AREA

Olfactory nerve (I)

OPTIC RECESS

TECTUM

Frontonasal process

Nasal cavity

SUBTHALAMUS

MESENCEPHALON

TEGMENTUM

Maxilla

Palatal process

Oral cavity

Sphenoid

HYPOTHALAMUS

Pituitary gland (anterior part)

Mandible

Tongue

Sella turcica

RHOMBENCEPHALON

PONS

UPPER MEDULLA

CEREBELLUM (HEMISPHERE)

Basal occipital

Hyoid bone?

Clavicle?

Rhombencephalic choroid plexus

RHOMBENCEPHALIC SUPERVENTRICLE (FUTURE FOURTH VENTRICLE)

Squamous occipital bone

Cervical vertebral column

Atlas spinal process?

Axis spinal process?

FONT KEY:
VENTRICULAR DIVISIONS - CAPITALS
Major brain structure - Times **Bold CAPITALS**
All other structures - Times Roman or **Bold**

PLATE 40C
CR 40 mm, GW 10.3, C6658
Sagittal
Slide 63, Section 1

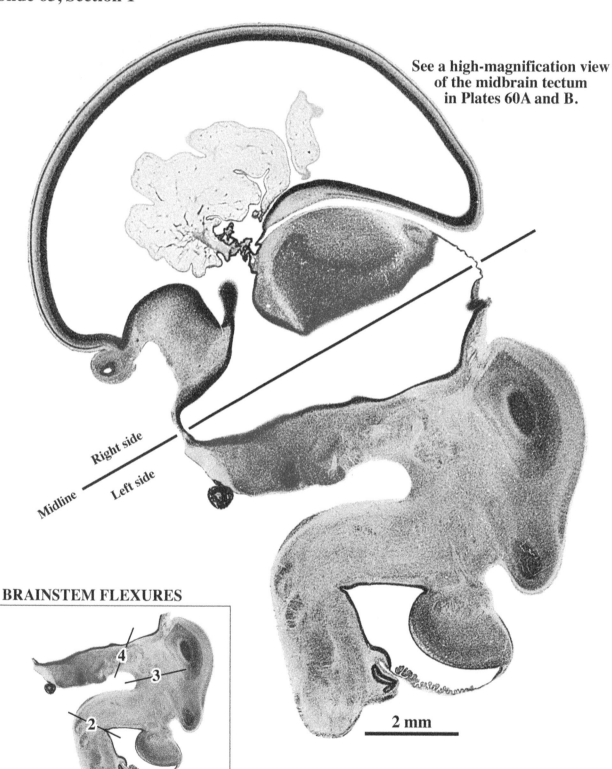

See a high-magnification view
of the midbrain tectum
in Plates 60A and B.

Right side

Midline   Left side

2 mm

BRAINSTEM FLEXURES

4
3
2

2. Pontine
3. Mesencephalic
4. Diencephalic

The head, major brain structures, and ventricular
divisions are labeled in Parts A and B of this plate on
the preceding pages.

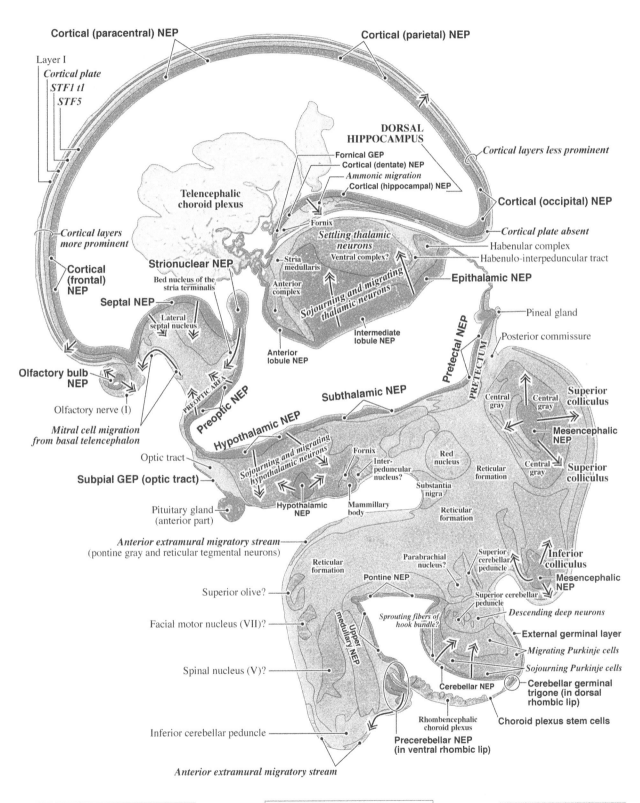

Cortical (paracentral) NEP

Cortical (parietal) NEP

Layer I
Cortical plate
STF1 t1
STF5

DORSAL HIPPOCAMPUS

Fornical GEP
Cortical (dentate) NEP
Ammonic migration
Cortical (hippocampal) NEP

Cortical layers less prominent

Telencephalic choroid plexus

Fornix
Settling thalamic neurons
Ventral complex?

Cortical (occipital) NEP

Cortical plate absent

Cortical layers more prominent

Strionuclear NEP

Stria medullaris

Cortical (frontal) NEP

Bed nucleus of the stria terminalis

Anterior complex

Sojourning and migrating thalamic neurons

Habenular complex
Habenulo-interpeduncular tract

Epithalamic NEP

Septal NEP

Lateral septal nucleus

Intermediate lobule NEP

Pineal gland

Pretectal NEP

Posterior commissure

Anterior lobule NEP

Olfactory bulb NEP

Olfactory nerve (I)

PRE

Central gray

Central gray

Superior colliculus

PRETECTUM

Mitral cell migration from basal telencephalon

Preoptic NEP

Hypothalamic NEP

Subthalamic NEP

Mesencephalic NEP

Central gray

Superior colliculus

Optic tract

Sojourning and migrating hypothalamic neurons

Fornix
Inter-peduncular nucleus?

Red nucleus

Reticular formation

Subpial GEP (optic tract)

Hypothalamic NEP

Mammillary body

Substantia nigra

Reticular formation

Pituitary gland (anterior part)

Anterior extramural migratory stream
(pontine gray and reticular tegmental neurons)

Reticular formation

Parabrachial nucleus?

Superior cerebellar peduncle

Inferior colliculus

Pontine NEP

Mesencephalic NEP

Superior olive?

Superior cerebellar peduncle

Descending deep neurons

Facial motor nucleus (VII)?

Upper medullary NEP

Sprouting fibers of hook bundle?

External germinal layer

Migrating Purkinje cells

Sojourning Purkinje cells

Spinal nucleus (V)?

Cerebellar NEP

Cerebellar germinal trigone (in dorsal rhombic lip)

Rhombencephalic choroid plexus

Choroid plexus stem cells

Inferior cerebellar peduncle

Precerebellar NEP (in ventral rhombic lip)

Anterior extramural migratory stream

FONT KEY:
Germinal zone - Helvetica bold
Transient structure - Times bold italic
Permanent structure - Times Roman or **Bold**

ABBREVIATIONS:
GEP - Glioepithelium
NEP - Neuroepithelium
STF - Stratified transitional field

Arrows indicate the presumed *direction of neuron migration* from neuroepithelial sources.

PLATE 41A
CR 40 mm, GW 10.3, C6658
Sagittal
Slide 67, Section 1

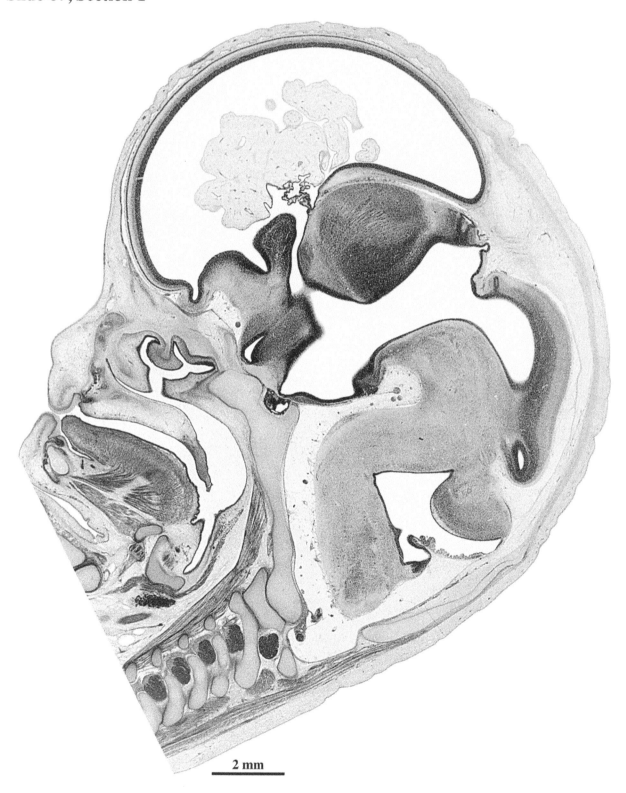

2 mm

Neuroepithelial divisions, glioepithelial divisions, and differentiating
structures are labeled in Parts C and D of this plate on the following pages.

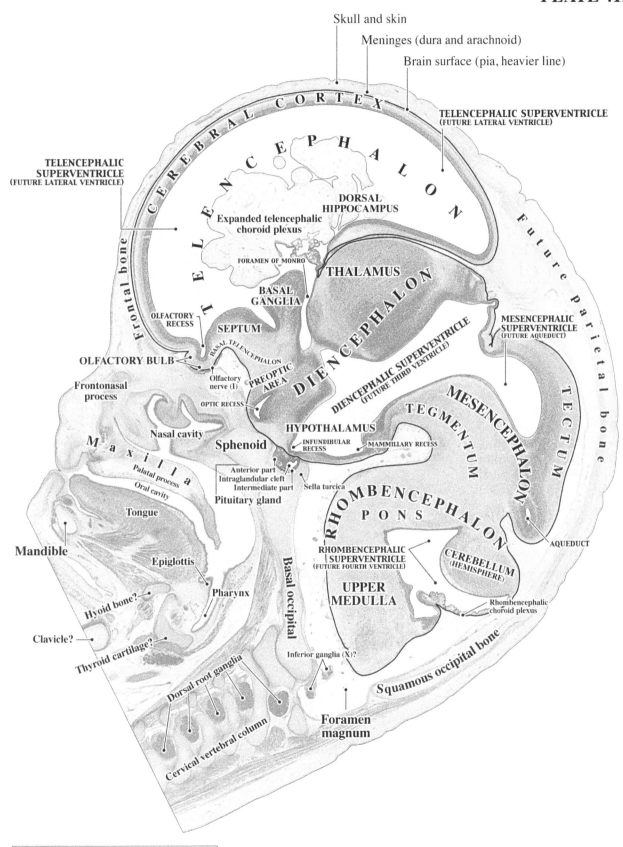

Skull and skin

Meninges (dura and arachnoid)

Brain surface (pia, heavier line)

CEREBRAL CORTEX

TELENCEPHALON

TELENCEPHALIC SUPERVENTRICLE
(FUTURE LATERAL VENTRICLE)

TELENCEPHALIC SUPERVENTRICLE
(FUTURE LATERAL VENTRICLE)

Future parietal bone

Frontal bone

DORSAL HIPPOCAMPUS

Expanded telencephalic choroid plexus

FORAMEN OF MONRO

THALAMUS

BASAL GANGLIA

OLFACTORY RECESS

SEPTUM

BASAL TELENCEPHALON

OLFACTORY BULB

Olfactory nerve (I)

PREOPTIC AREA

DIENCEPHALON

DIENCEPHALIC SUPERVENTRICLE
(FUTURE THIRD VENTRICLE)

MESENCEPHALIC SUPERVENTRICLE
(FUTURE AQUEDUCT)

Frontonasal process

OPTIC RECESS

Nasal cavity

M a x i l l a

Sphenoid

HYPOTHALAMUS

INFUNDIBULAR RECESS

MAMMILLARY RECESS

TEGMENTUM

MESENCEPHALON

TECTUM

Palatal process

Oral cavity

Anterior part
Intraglandular cleft
Intermediate part
Pituitary gland

Sella turcica

Tongue

RHOMBENCEPHALON

P O N S

AQUEDUCT

Mandible

Epiglottis

Pharynx

RHOMBENCEPHALIC SUPERVENTRICLE
(FUTURE FOURTH VENTRICLE)

CEREBELLUM
(HEMISPHERE)

Hyoid bone?

Basal occipital

UPPER MEDULLA

Rhombencephalic choroid plexus

Clavicle?

Thyroid cartilage?

Inferior ganglia (X)?

Dorsal root ganglia

Foramen magnum

Squamous occipital bone

Cervical vertebral column

FONT KEY:
VENTRICULAR DIVISIONS – CAPITALS
Major brain structure - Times **Bold CAPITALS**
All other structures - Times Roman or **Bold**

PLATE 41C
CR 40 mm, GW 10.3, C6658
Sagittal
Slide 67, Section 1

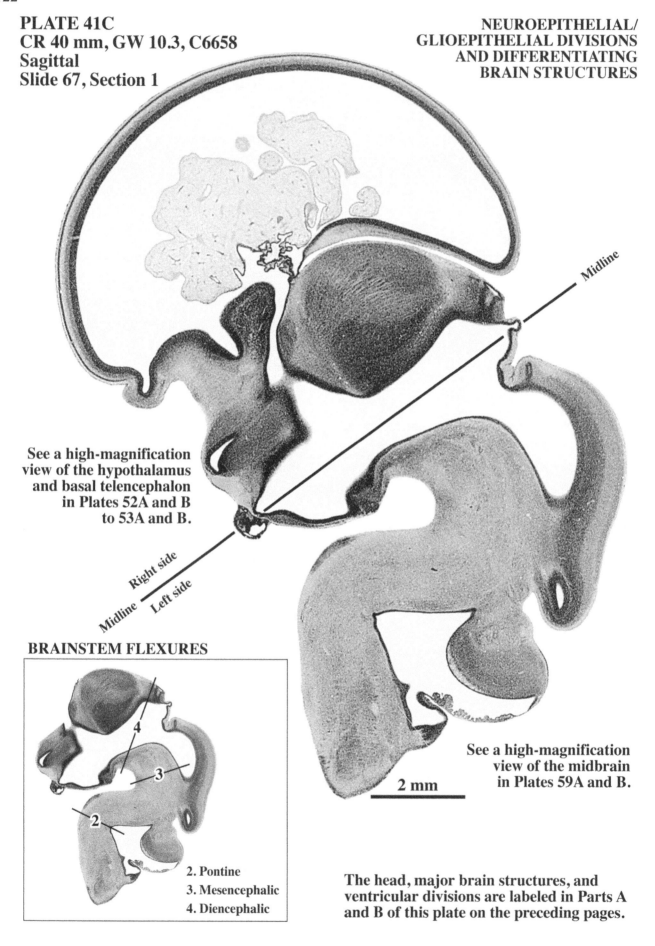

Midline

**See a high-magnification
view of the hypothalamus
and basal telencephalon
in Plates 52A and B
to 53A and B.**

Right side

Midline   Left side

BRAINSTEM FLEXURES

4

3

2

2. Pontine

3. Mesencephalic

4. Diencephalic

**See a high-magnification
view of the midbrain
in Plates 59A and B.**

2 mm

**The head, major brain structures, and
ventricular divisions are labeled in Parts A
and B of this plate on the preceding pages.**

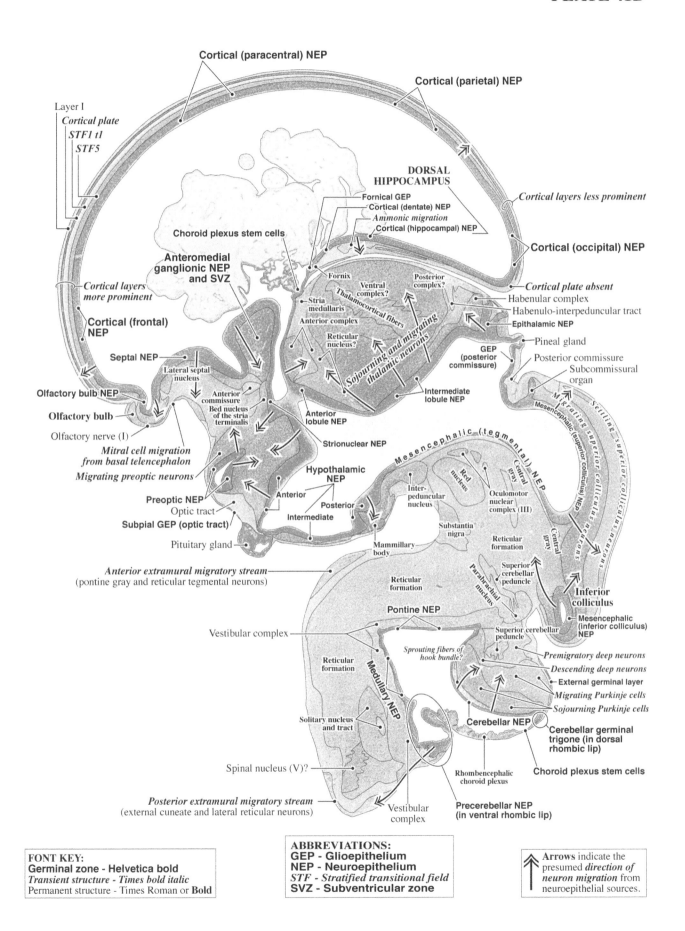

Cortical (paracentral) NEP

Cortical (parietal) NEP

Layer I
Cortical plate
STF1 t1
STF5

**DORSAL
HIPPOCAMPUS**

Fornical GEP
Cortical (dentate) NEP
Ammonic migration
Cortical (hippocampal) NEP

Cortical layers less prominent

Choroid plexus stem cells

**Anteromedial
ganglionic NEP
and SVZ**

Fornix

Posterior
complex?

Cortical (occipital) NEP

*Cortical layers
more prominent*

Ventral
complex?

Thalamocortical fibers

Cortical plate absent
Habenular complex
Habenulo-interpeduncular tract

Stria
medullaris

Anterior complex

Epithalamic NEP

**Cortical (frontal)
NEP**

Reticular
nucleus?

*Sojourning and migrating
thalamic neurons*

Pineal gland

GEP
(posterior
commissure)

Posterior commissure
Subcommissural
organ

Septal NEP

Lateral septal
nucleus

Intermediate
lobule NEP

Mesencephalic (superior colliculus) NEP

Migrating superior colliculus neurons

Olfactory bulb NEP

Anterior
commissure
Bed nucleus
of the stria
terminalis

Settling superior colliculus neurons

Olfactory bulb

Anterior
lobule NEP

Mesencephalic (tegmental) NEP

Olfactory nerve (I)

Strionuclear NEP

Red
nucleus

Central
gray

*Mitral cell migration
from basal telencephalon*

Inter-
peduncular
nucleus

Oculomotor
nuclear
complex (III)

Migrating preoptic neurons

**Hypothalamic
NEP**

Central
gray

Preoptic NEP

Anterior

Substantia
nigra

Reticular
formation

Optic tract

Posterior

Subpial GEP (optic tract)

Intermediate

Mammillary
body

Reticular
formation

Superior
cerebellar
peduncle

**Inferior
colliculus**

Pituitary gland

Anterior extramural migratory stream
(pontine gray and reticular tegmental neurons)

Reticular
formation

Parabrachial
nucleus

Mesencephalic
(inferior colliculus)
NEP

Vestibular complex

Pontine NEP

Superior cerebellar
peduncle

Premigratory deep neurons

Descending deep neurons

*Sprouting fibers of
hook bundle?*

External germinal layer

Medullary NEP

Migrating Purkinje cells

Reticular
formation

Sojourning Purkinje cells

Solitary nucleus
and tract

Cerebellar NEP

**Cerebellar germinal
trigone (in dorsal
rhombic lip)**

Spinal nucleus (V)?

Rhombencephalic
choroid plexus

Choroid plexus stem cells

Posterior extramural migratory stream
(external cuneate and lateral reticular neurons)

Vestibular
complex

**Precerebellar NEP
(in ventral rhombic lip)**

FONT KEY:
Germinal zone - **Helvetica bold**
Transient structure - *Times bold italic*
Permanent structure - Times Roman or **Bold**

ABBREVIATIONS:
GEP - Glioepithelium
NEP - Neuroepithelium
STF - *Stratified transitional field*
SVZ - Subventricular zone

Arrows indicate the
presumed *direction of
neuron migration* from
neuroepithelial sources.

PLATE 42A
CR 40 mm, GW 10.3, C6658
Sagittal
Slide 71, Section 2

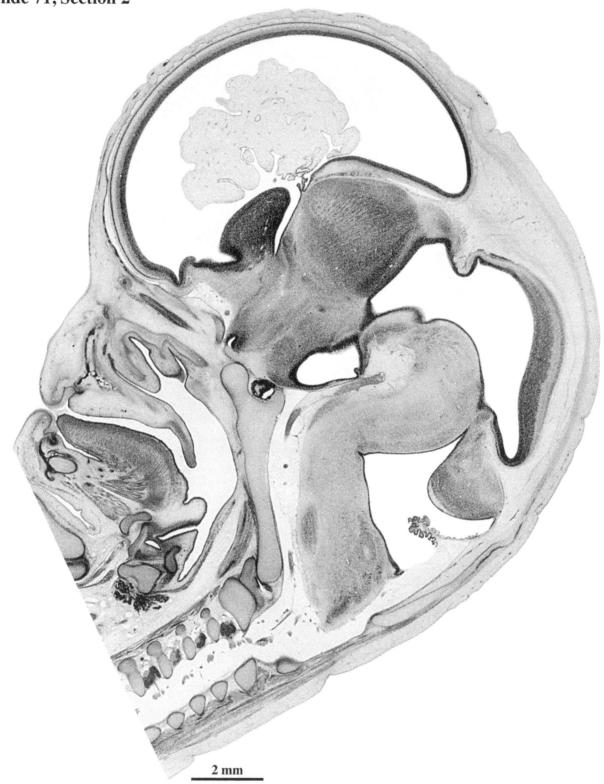

2 mm

Neuroepithelial divisions, glioepithelial divisions, and differentiating
structures are labeled in Parts C and D of this plate on the following pages.

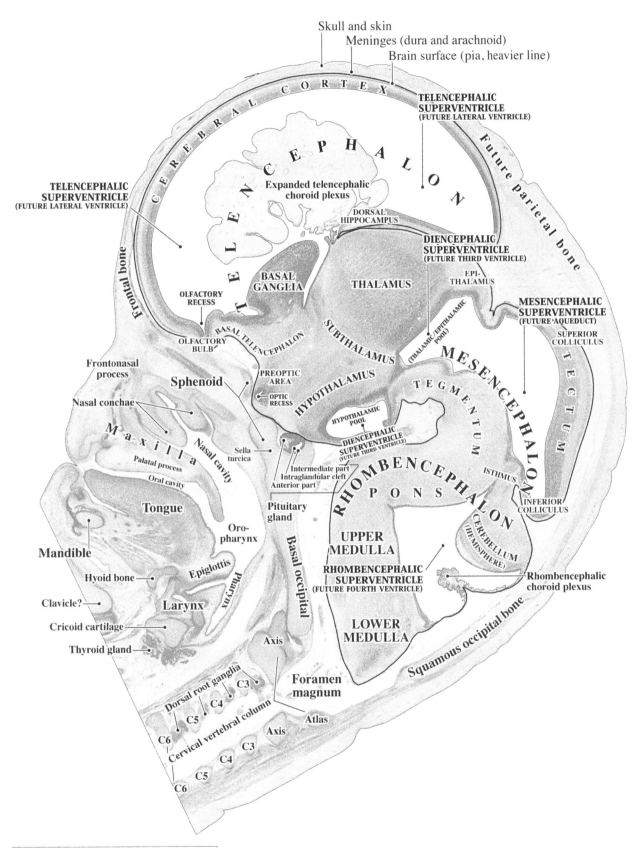

Skull and skin
Meninges (dura and arachnoid)
Brain surface (pia, heavier line)

TELENCEPHALIC
SUPERVENTRICLE
(FUTURE LATERAL VENTRICLE)

CEREBRAL CORTEX

Future parietal bone

TELENCEPHALON

Expanded telencephalic
choroid plexus

DORSAL
HIPPOCAMPUS

TELENCEPHALIC
SUPERVENTRICLE
(FUTURE LATERAL VENTRICLE)

DIENCEPHALIC
SUPERVENTRICLE
(FUTURE THIRD VENTRICLE)

Frontal bone

THALAMUS

EPI-
THALAMUS

MESENCEPHALIC
SUPERVENTRICLE
(FUTURE AQUEDUCT)

BASAL
GANGLIA

OLFACTORY
RECESS

SUBTHALAMUS

(THALAMIC EPITHALAMIC
POOL)

SUPERIOR
COLLICULUS

OLFACTORY
BULB

BASAL TELENCEPHALON

HYPOTHALAMUS

MESENCEPHALON

TECTUM

*Frontonasal
process*

PREOPTIC
AREA

TEGMENTUM

Nasal conchae

Sphenoid

OPTIC
RECESS

HYPOTHALAMIC
POOL

Maxilla

Nasal cavity

DIENCEPHALIC
SUPERVENTRICLE
(FUTURE THIRD VENTRICLE)

ISTHMUS

Palatal process

*Sella
turcica*

INFERIOR
COLLICULUS

Oral cavity

Intermediate part
Intraglandular cleft
Anterior part

RHOMBENCEPHALON

PONS

Tongue

**Pituitary
gland**

CEREBELLUM
(HEMISPHERE)

Mandible

*Oro-
pharynx*

UPPER
MEDULLA

Epiglottis

Hyoid bone

Pharynx

RHOMBENCEPHALIC
SUPERVENTRICLE
(FUTURE FOURTH VENTRICLE)

Rhombencephalic
choroid plexus

Clavicle?

Larynx

Cricoid cartilage

Basal occipital

LOWER
MEDULLA

Thyroid gland

Axis

Squamous occipital bone

Dorsal root ganglia

C3

**Foramen
magnum**

C4

C5

Atlas

C6

Cervical vertebral column

Axis

C3

C4

C5

C6

FONT KEY:
VENTRICULAR DIVISIONS - CAPITALS
Major brain structure - Times **Bold CAPITALS**
All other structures - Times Roman or **Bold**

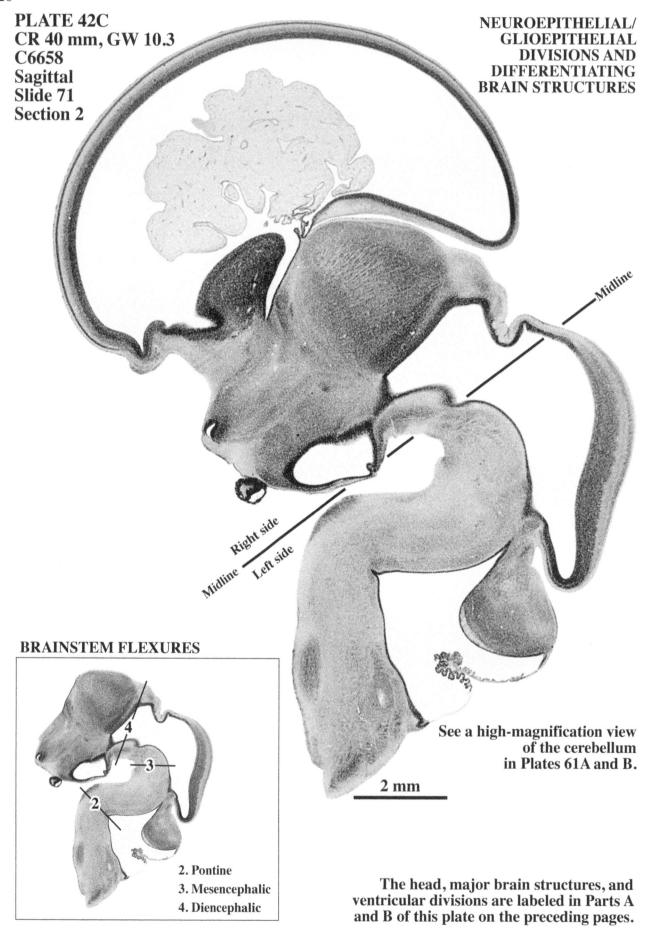

PLATE 42C
CR 40 mm, GW 10.3
C6658
Sagittal
Slide 71
Section 2

NEUROEPITHELIAL/
GLIOEPITHELIAL
DIVISIONS AND
DIFFERENTIATING
BRAIN STRUCTURES

Midline

Right side

Midline   Left side

BRAINSTEM FLEXURES

4

3

2

2. Pontine
3. Mesencephalic
4. Diencephalic

See a high-magnification view
of the cerebellum
in Plates 61A and B.

2 mm

The head, major brain structures, and
ventricular divisions are labeled in Parts A
and B of this plate on the preceding pages.

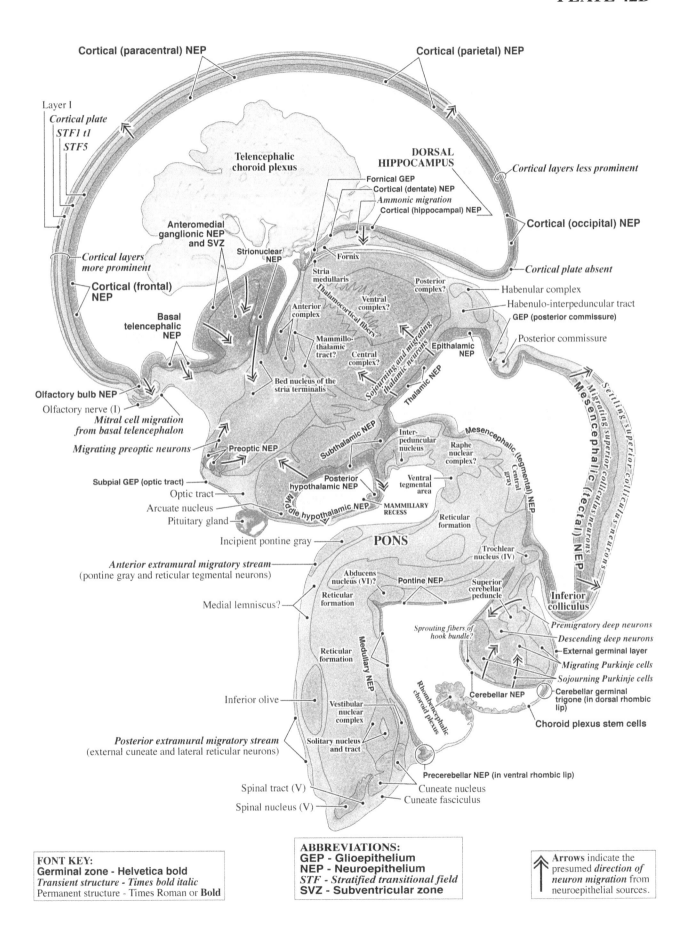

Cortical (paracentral) NEP

Cortical (parietal) NEP

Layer I
Cortical plate
STF1 t1
STF5

Telencephalic
choroid plexus

**DORSAL
HIPPOCAMPUS**

Fornical GEP
Cortical (dentate) NEP
Ammonic migration
Cortical (hippocampal) NEP

Cortical layers less prominent

Cortical (occipital) NEP

Cortical plate absent

Anteromedial
ganglionic NEP
and SVZ

Strionuclear
NEP

Fornix

Stria
medullaris

Thalamocortical fibers

Ventral
complex?

Anterior
complex

Posterior
complex?

Habenular complex

Habenulo-interpeduncular tract

GEP (posterior commissure)

Posterior commissure

*Cortical layers
more prominent*

**Cortical (frontal)
NEP**

**Basal
telencephalic
NEP**

Mammillo-
thalamic
tract?

Central
complex?

Bed nucleus of the
stria terminalis

*Sojourning and migrating
thalamic neurons*

Epithalamic
NEP

Thalamic NEP

Settling superior colliculus neurons

Migrating superior colliculus neurons

Olfactory bulb NEP

Olfactory nerve (I)

*Mitral cell migration
from basal telencephalon*

Migrating preoptic neurons

Preoptic NEP

Subthalamic NEP

Inter-
peduncular
nucleus

Raphe
nuclear
complex?

Mesencephalic (tegmental) NEP

*Central
gray*

Mesencephalic (tectal) NEP

Subpial GEP (optic tract)

Optic tract

Arcuate nucleus

Pituitary gland

**Posterior
hypothalamic NEP**

Middle hypothalamic NEP

Ventral
tegmental
area

**MAMMILLARY
RECESS**

Reticular
formation

Incipient pontine gray

PONS

Trochlear
nucleus (IV)

**Inferior
colliculus**

Anterior extramural migratory stream
(pontine gray and reticular tegmental neurons)

Abducens
nucleus (VI)?

Pontine NEP

Superior
cerebellar
peduncle

Premigratory deep neurons

Descending deep neurons

Medial lemniscus?

Reticular
formation

*Sprouting fibers of
hook bundle?*

External germinal layer

Migrating Purkinje cells

Sojourning Purkinje cells

Reticular
formation

Medullary NEP

Cerebellar NEP

Cerebellar germinal
trigone (in dorsal rhombic
lip)

Inferior olive

Vestibular
nuclear
complex

*Rhombencephalic
choroid plexus*

Choroid plexus stem cells

Posterior extramural migratory stream
(external cuneate and lateral reticular neurons)

Solitary nucleus
and tract

Precerebellar NEP (in ventral rhombic lip)

Cuneate nucleus

Cuneate fasciculus

Spinal tract (V)

Spinal nucleus (V)

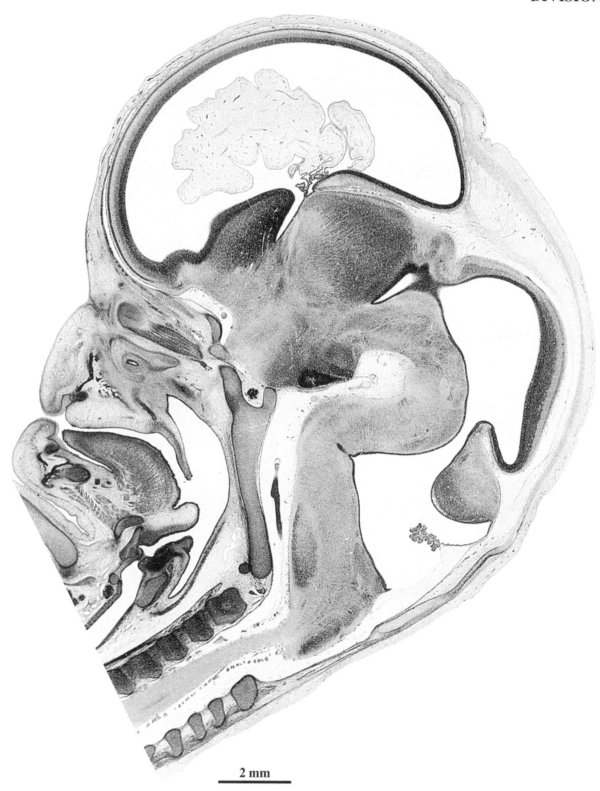

2 mm

Neuroepithelial divisions, glioepithelial divisions, and differentiating
structures are labeled in Parts C and D of this plate on the following pages.

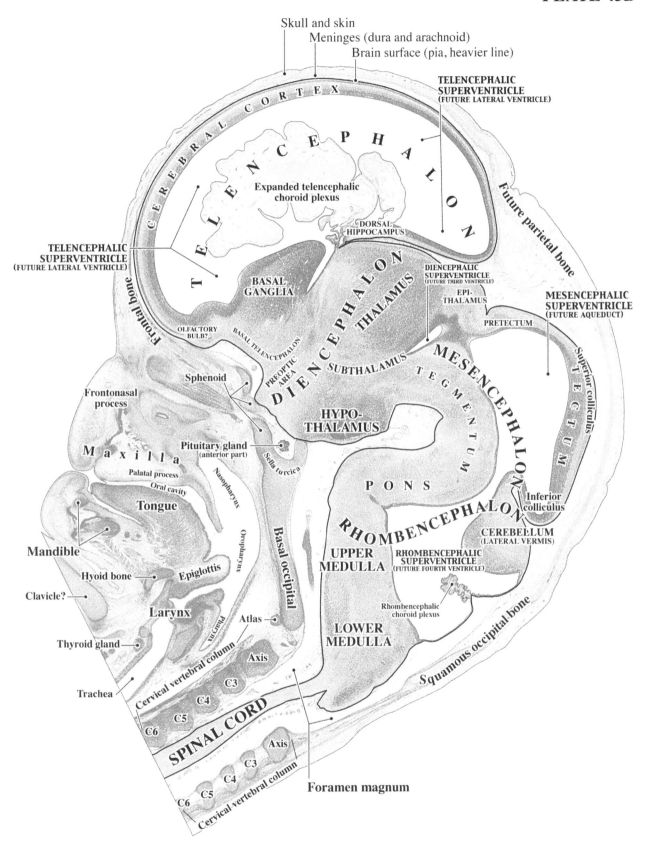

Skull and skin
Meninges (dura and arachnoid)
Brain surface (pia, heavier line)

**TELENCEPHALIC
SUPERVENTRICLE
(FUTURE LATERAL VENTRICLE)**

CEREBRAL CORTEX

TELENCEPHALON

Expanded telencephalic
choroid plexus

**DORSAL
HIPPOCAMPUS**

Future parietal bone

**TELENCEPHALIC
SUPERVENTRICLE
(FUTURE LATERAL VENTRICLE)**

**BASAL
GANGLIA**

DIENCEPHALON

THALAMUS

**DIENCEPHALIC
SUPERVENTRICLE
(FUTURE THIRD VENTRICLE)**

**EPI-
THALAMUS**

**MESENCEPHALIC
SUPERVENTRICLE
(FUTURE AQUEDUCT)**

OLFACTORY
BULB?

BASAL TELENCEPHALON

PRECTUM

SUBTHALAMUS

MESENCEPHALON

PREOPTIC
AREA

TECTUM

Frontal bone

Frontonasal
process

Sphenoid

**HYPO-
THALAMUS**

T E G M E N T U M

Superior colliculus

Pituitary gland
(anterior part)

Sella turcica

Nasopharynx

Maxilla

Palatal process

Oral cavity

P O N S

Inferior
colliculus

Tongue

Oropharynx

RHOMBENCEPHALON

**CEREBELLUM
(LATERAL VERMIS)**

Mandible

**UPPER
MEDULLA**

**RHOMBENCEPHALIC
SUPERVENTRICLE
(FUTURE FOURTH VENTRICLE)**

Hyoid bone

Epiglottis

Clavicle?

Rhombencephalic
choroid plexus

Larynx

Pharynx

Atlas

Basal occipital

**LOWER
MEDULLA**

Thyroid gland

Squamous occipital bone

Axis

Trachea

C3

Cervical vertebral column

C4

C5

C6

SPINAL CORD

Axis

C3

C4

C5

C6

Cervical vertebral column

Foramen magnum

FONT KEY:
VENTRICULAR DIVISIONS – CAPITALS
Major brain structure - Times **Bold CAPITALS**
All other structures - Times Roman or **Bold**

PLATE 43C

**NEUROEPITHELIAL/
GLIOEPITHELIAL
DIVISIONS AND
DIFFERENTIATING
BRAIN
STRUCTURES**

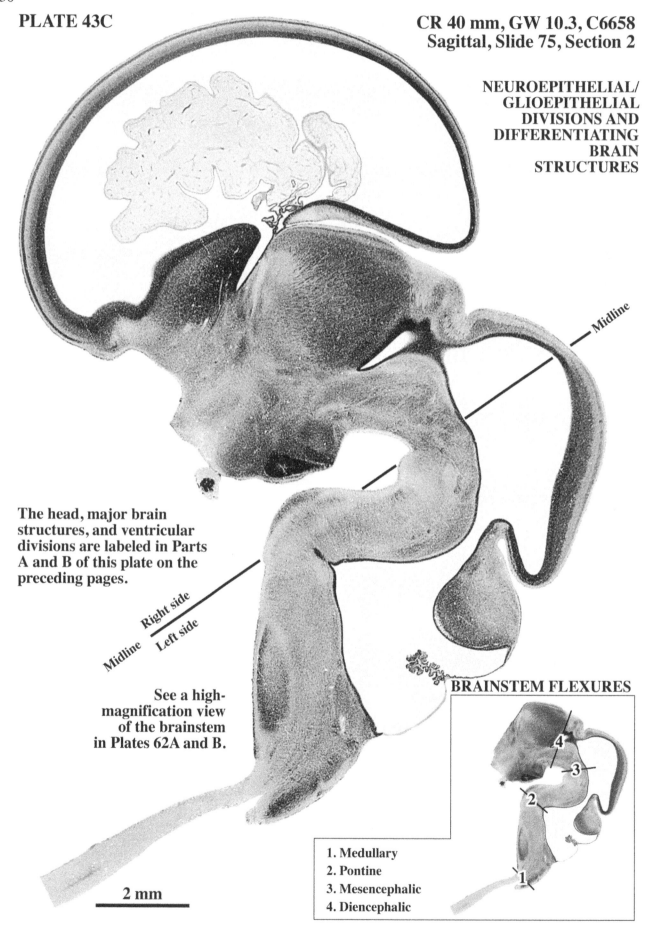

Midline

The head, major brain
structures, and ventricular
divisions are labeled in Parts
A and B of this plate on the
preceding pages.

Midline Right side Left side

See a high-
magnification view
of the brainstem
in Plates 62A and B.

2 mm

BRAINSTEM FLEXURES

1. Medullary
2. Pontine
3. Mesencephalic
4. Diencephalic

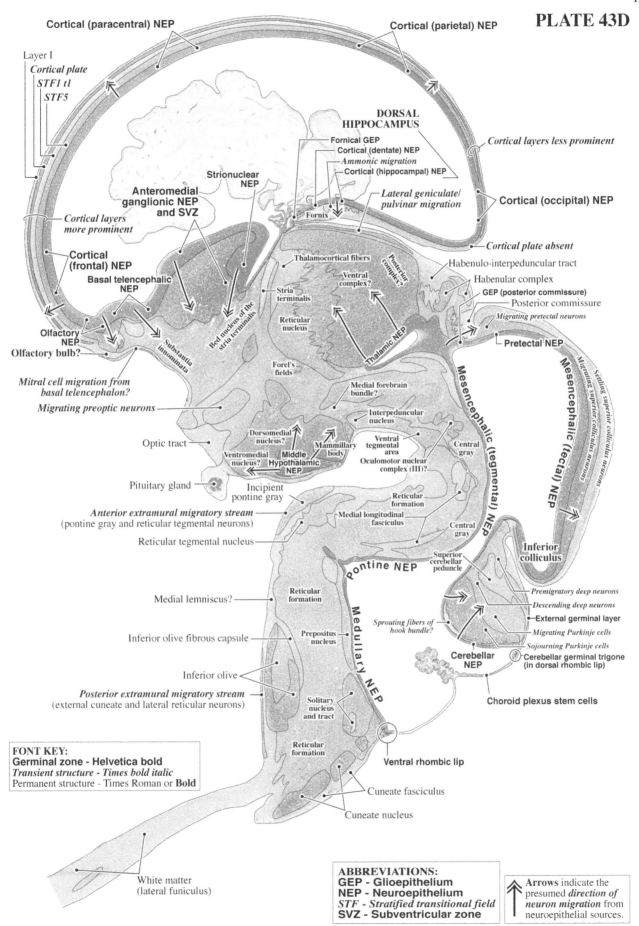

Cortical (paracentral) NEP

Cortical (parietal) NEP

Layer I
Cortical plate
STF1 t1
STF5

**DORSAL
HIPPOCAMPUS**

Fornical GEP
Cortical (dentate) NEP
Ammonic migration
Cortical (hippocampal) NEP

Cortical layers less prominent

Strionuclear
NEP

**Anteromedial
ganglionic NEP
and SVZ**

*Lateral geniculate/
pulvinar migration*

Cortical (occipital) NEP

Fornix

*Cortical layers
more prominent*

**Cortical
(frontal) NEP**

**Basal telencephalic
NEP**

Thalamocortical fibers

*Posterior
complex?*

*Ventral
complex?*

Cortical plate absent

Habenulo-interpeduncular tract

Habenular complex

GEP (posterior commissure)

Posterior commissure

Migrating pretectal neurons

Stria
terminalis

**Olfactory
NEP**

Olfactory bulb?

*Substantia
innominata*

Reticular
nucleus

Pretectal NEP

Bed nucleus of the
stria terminalis

Thalamic NEP

*Mitral cell migration from
basal telencephalon?*

Migrating preoptic neurons

Forel's
fields

Medial forebrain
bundle?

Interpeduncular
nucleus

Central
gray

Optic tract

*Dorsomedial
nucleus?*

Ventral
tegmental
area

Oculomotor nuclear
complex (III)?

*Ventromedial
nucleus?* **Middle
Hypothalamic
NEP**

Mammillary
body

Mesencephalic (tegmental) NEP

Mesencephalic (tectal) NEP

Settling superior colliculus neurons

Migrating superior colliculus neurons

Pituitary gland

*Incipient
pontine gray*

Central
gray

Reticular
formation

Anterior extramural migratory stream
(pontine gray and reticular tegmental neurons)

Medial longitudinal
fasciculus

Reticular tegmental nucleus

**Inferior
colliculus**

Superior
cerebellar
peduncle

Premigratory deep neurons

Medial lemniscus?

Reticular
formation

Pontine NEP

Descending deep neurons

External germinal layer

*Sprouting fibers of
hook bundle?*

Migrating Purkinje cells

Inferior olive fibrous capsule

*Prepositus
nucleus*

Medullary NEP

Sojourning Purkinje cells

**Cerebellar
NEP**

Cerebellar germinal trigone
(in dorsal rhombic lip)

Inferior olive

Choroid plexus stem cells

Posterior extramural migratory stream
(external cuneate and lateral reticular neurons)

*Solitary
nucleus
and tract*

Reticular
formation

Ventral rhombic lip

FONT KEY:
Germinal zone - Helvetica bold
Transient structure - Times bold italic
Permanent structure - Times Roman or **Bold**

Cuneate fasciculus

Cuneate nucleus

White matter
(lateral funiculus)

ABBREVIATIONS:
GEP - Glioepithelium
NEP - Neuroepithelium
STF - Stratified transitional field
SVZ - Subventricular zone

Arrows indicate the
presumed *direction of
neuron migration* from
neuroepithelial sources.

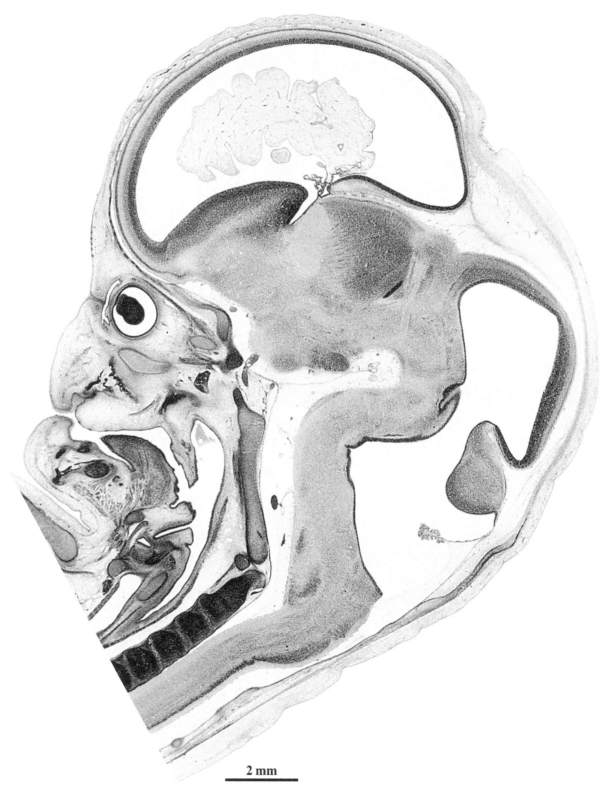

2 mm

Neuroepithelial divisions, glioepithelial divisions, and differentiating
structures are labeled in Parts C and D of this plate on the following pages.

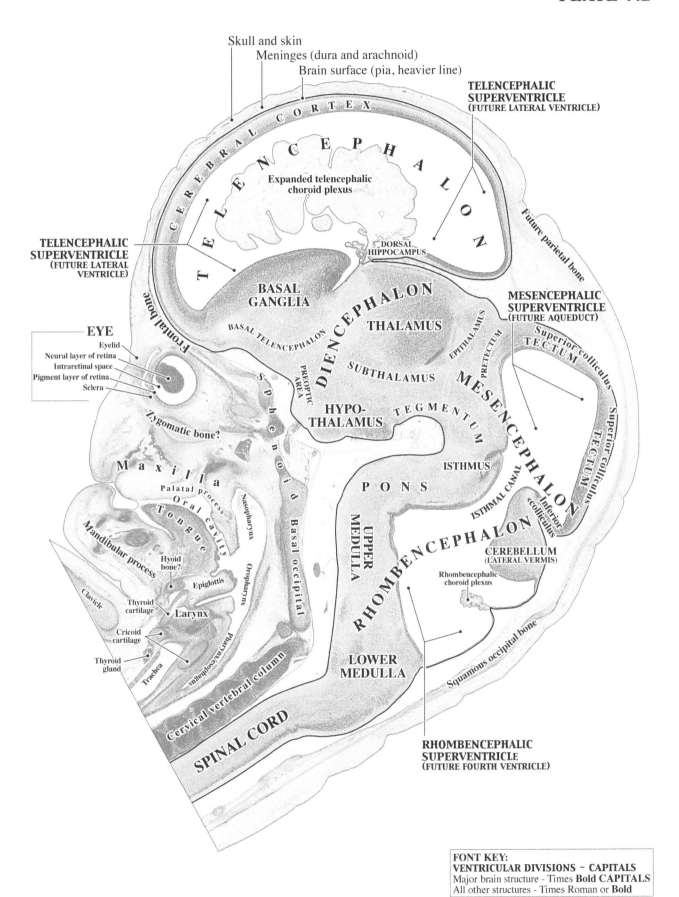

Skull and skin
Meninges (dura and arachnoid)
Brain surface (pia, heavier line)

**TELENCEPHALIC
SUPERVENTRICLE
(FUTURE LATERAL VENTRICLE)**

C E R E B R A L   C O R T E X

T E L E N C E P H A L O N

**Expanded telencephalic
choroid plexus**

Future parietal bone

**DORSAL
HIPPOCAMPUS**

**TELENCEPHALIC
SUPERVENTRICLE
(FUTURE LATERAL
VENTRICLE)**

**BASAL
GANGLIA**

DIENCEPHALON

BASAL TELENCEPHALON

THALAMUS

EPITHALAMUS

PRETECTUM

**MESENCEPHALIC
SUPERVENTRICLE
(FUTURE AQUEDUCT)**

Superior colliculus
TECTUM

EYE

Eyelid
Neural layer of retina
Intraretinal space
Pigment layer of retina
Sclera

Frontal bone

PREOPTIC
AREA

SUBTHALAMUS

**HYPO-
THALAMUS**

TEGMENTUM

MESENCEPHALON

Superior colliculus
TECTUM

Zygomatic bone?

S
p
h
e
n
o
i
d

ISTHMUS

ISTHMAL CANAL

Inferior
colliculus

M a x i l l a

Palatal process

Oral cavity

T o n g u e

Nasopharynx

B
a
s
a
l

o
c
c
i
p
i
t
a
l

P O N S

**UPPER
MEDULLA**

RHOMBENCEPHALON

**CEREBELLUM
(LATERAL VERMIS)**

Mandibular process

Hyoid
bone?

Oropharynx

**Rhombencephalic
choroid plexus**

Clavicle

Epiglottis

Thyroid
cartilage

Larynx

Cricoid
cartilage

Pharyngoesophagus

**LOWER
MEDULLA**

Squamous occipital bone

Thyroid
gland

Trachea

Cervical vertebral column

SPINAL CORD

**RHOMBENCEPHALIC
SUPERVENTRICLE
(FUTURE FOURTH VENTRICLE)**

**FONT KEY:
VENTRICULAR DIVISIONS – CAPITALS**
Major brain structure - Times **Bold CAPITALS**
All other structures - Times Roman or **Bold**

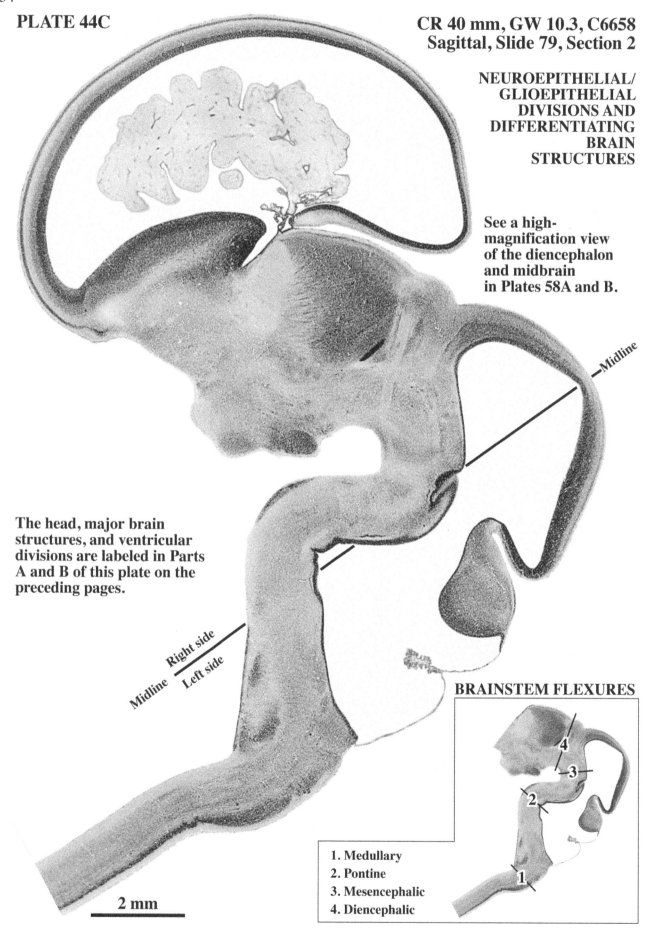

134

CR 40 mm, GW 10.3, C6658
Sagittal, Slide 79, Section 2

NEUROEPITHELIAL/
GLIOEPITHELIAL
DIVISIONS AND
DIFFERENTIATING
BRAIN
STRUCTURES

See a high-
magnification view
of the diencephalon
and midbrain
in Plates 58A and B.

Midline

The head, major brain
structures, and ventricular
divisions are labeled in Parts
A and B of this plate on the
preceding pages.

Right side

Left side

Midline

BRAINSTEM FLEXURES

4

3

2

1

1. Medullary
2. Pontine
3. Mesencephalic
4. Diencephalic

2 mm

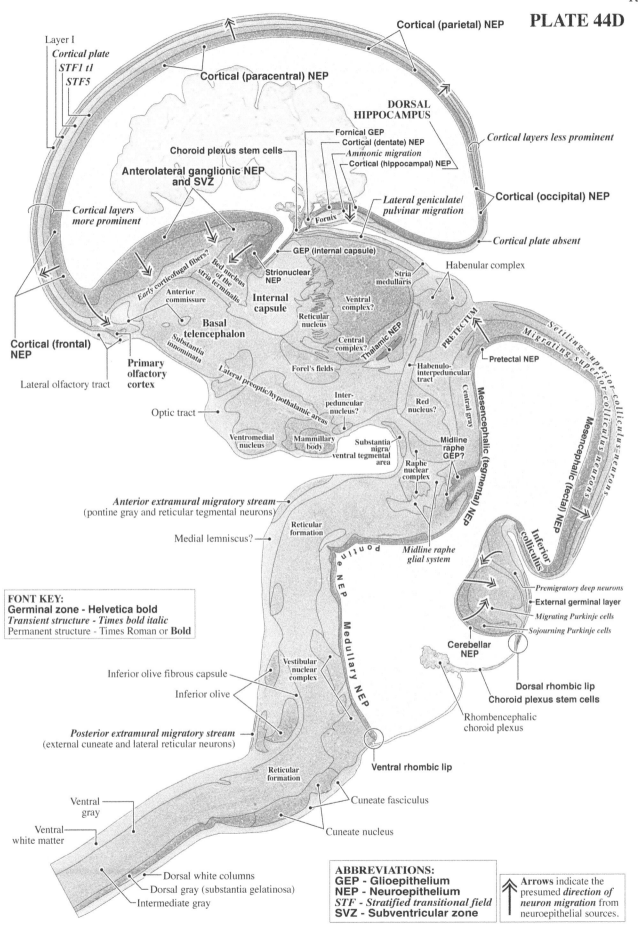

PLATE 44D

Layer I
Cortical plate
STF1 tl
STF5

Cortical (parietal) NEP

Cortical (paracentral) NEP

**DORSAL
HIPPOCAMPUS**

Choroid plexus stem cells

Fornical GEP
Cortical (dentate) NEP
Ammonic migration
Cortical (hippocampal) NEP

Cortical layers less prominent

**Anterolateral ganglionic NEP
and SVZ**

*Lateral geniculate/
pulvinar migration*

Cortical (occipital) NEP

*Cortical layers
more prominent*

Fornix

GEP (internal capsule)

Cortical plate absent

**Cortical (frontal)
NEP**

Early corticofugal fibers?

*Bed nucleus
of the
stria terminalis*

Strionuclear
NEP

Stria
medullaris

Habenular complex

Anterior
commissure

**Internal
capsule**

Ventral
complex?

Stria
medullaris

**Basal
telencephalon**

Reticular
nucleus

Thalamic NEP

PRETECTUM

Settling superior colliculus neurons

Migrating superior colliculus neurons

*Substantia
innominata*

Central
complex?

Pretectal NEP

**Primary
olfactory
cortex**

Lateral preoptic/hypothalamic areas

Forel's fields

Habenulo-
interpeduncular
tract

Lateral olfactory tract

Optic tract

Inter-
peduncular
nucleus?

Red
nucleus?

*Central
gray*

Mesencephalic (tegmental) NEP

Mesencephalic (tectal) NEP

Ventromedial
nucleus

Mammillary
body

Substantia
nigra/
ventral tegmental
area

Midline
raphe
GEP?

Raphe
nuclear
complex

Anterior extramural migratory stream
(pontine gray and reticular tegmental neurons)

Reticular
formation

*Midline raphe
glial system*

*Inferior
colliculus*

Medial lemniscus?

Pontine NEP

Premigratory deep neurons
External germinal layer
Migrating Purkinje cells
Sojourning Purkinje cells

FONT KEY:
Germinal zone - Helvetica bold
Transient structure - Times bold italic
Permanent structure - Times Roman or **Bold**

Medullary NEP

Vestibular
nuclear
complex

**Cerebellar
NEP**

Dorsal rhombic lip
Choroid plexus stem cells

Inferior olive fibrous capsule

Inferior olive

Rhombencephalic
choroid plexus

Posterior extramural migratory stream
(external cuneate and lateral reticular neurons)

Reticular
formation

Ventral rhombic lip

Ventral
gray

Cuneate fasciculus

Ventral
white matter

Cuneate nucleus

Dorsal white columns
Dorsal gray (substantia gelatinosa)
Intermediate gray

ABBREVIATIONS:
GEP - Glioepithelium
NEP - Neuroepithelium
STF - Stratified transitional field
SVZ - Subventricular zone

Arrows indicate the
presumed *direction of
neuron migration* from
neuroepithelial sources.

PLATE 45A
CR 40 mm, GW 10.3, C6658
Sagittal, Slide 83, Section 1

HEAD STRUCTURES,
MAJOR BRAIN REGIONS,
AND VENTRICULAR
DIVISIONS

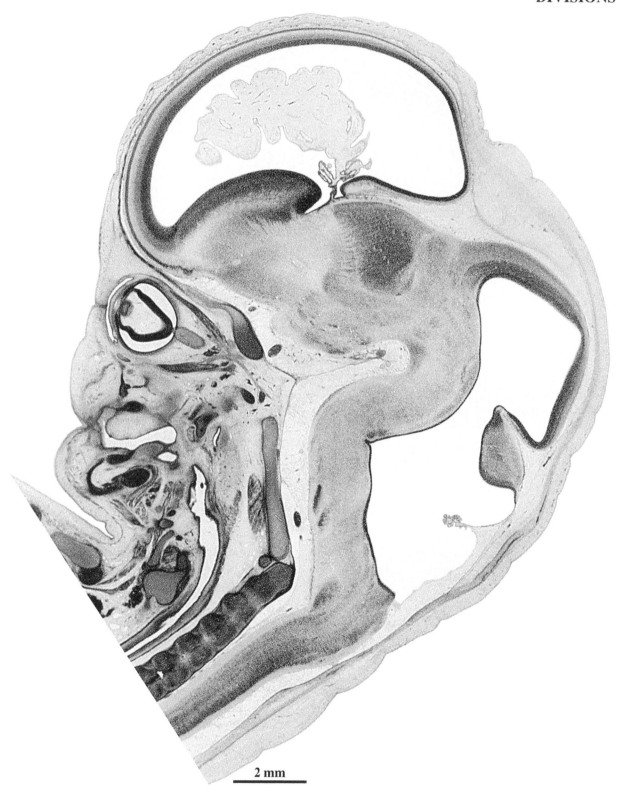

2 mm

Neuroepithelial divisions, glioepithelial divisions, and differentiating
structures are labeled in Parts C and D of this plate on the following pages.

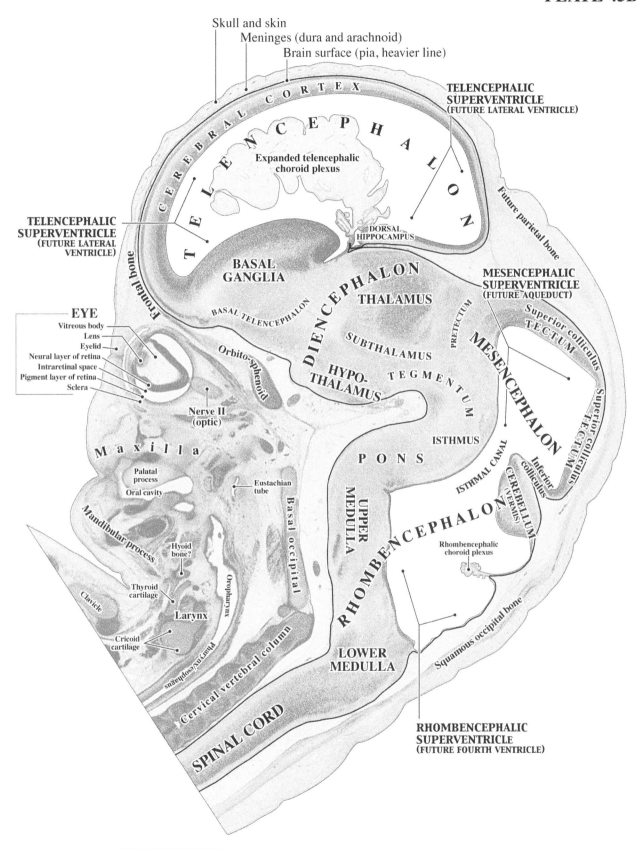

Skull and skin
Meninges (dura and arachnoid)
Brain surface (pia, heavier line)

C E R E B R A L   C O R T E X

**TELENCEPHALIC
SUPERVENTRICLE
(FUTURE LATERAL VENTRICLE)**

T E L E N C E P H A L O N

Expanded telencephalic
choroid plexus

Future parietal bone

**DORSAL
HIPPOCAMPUS**

**TELENCEPHALIC
SUPERVENTRICLE
(FUTURE LATERAL
VENTRICLE)**

Frontal bone

**BASAL
GANGLIA**

DIENCEPHALON

THALAMUS

**MESENCEPHALIC
SUPERVENTRICLE
(FUTURE AQUEDUCT)**

BASAL TELENCEPHALON

SUBTHALAMUS

PRETECTUM

Superior colliculus
TECTUM

MESENCEPHALON

EYE

Vitreous body
Lens
Eyelid
Neural layer of retina
Intraretinal space
Pigment layer of retina
Sclera

Orbito-sphenoid

**HYPO-
THALAMUS**

T E G M E N T U M

Superior colliculus
TECTUM

Nerve II
(optic)

M a x i l l a

ISTHMUS

P O N S

Inferior
colliculus

**CEREBELLUM
(VERMIS)**

Palatal
process

Oral cavity

Eustachian
tube

**UPPER
MEDULLA**

ISTHMAL CANAL

Mandibular process

Basal occipital

RHOMBENCEPHALON

Rhombencephalic
choroid plexus

Hyoid
bone?

Oropharynx

Thyroid
cartilage

Clavicle

Larynx

Laryngopharynx

Cervical vertebral column

Squamous occipital bone

Cricoid
cartilage

**LOWER
MEDULLA**

SPINAL CORD

**RHOMBENCEPHALIC
SUPERVENTRICLE
(FUTURE FOURTH VENTRICLE)**

FONT KEY:
VENTRICULAR DIVISIONS − CAPITALS
Major brain structure - Times **Bold CAPITALS**
All other structures - Times Roman or **Bold**

PLATE 45C

**NEUROEPITHELIAL/
GLIOEPITHELIAL
DIVISIONS AND
DIFFERENTIATING
BRAIN
STRUCTURES**

See a high-magnification
view of the lateral
forebrain in
Plates 55A and B.

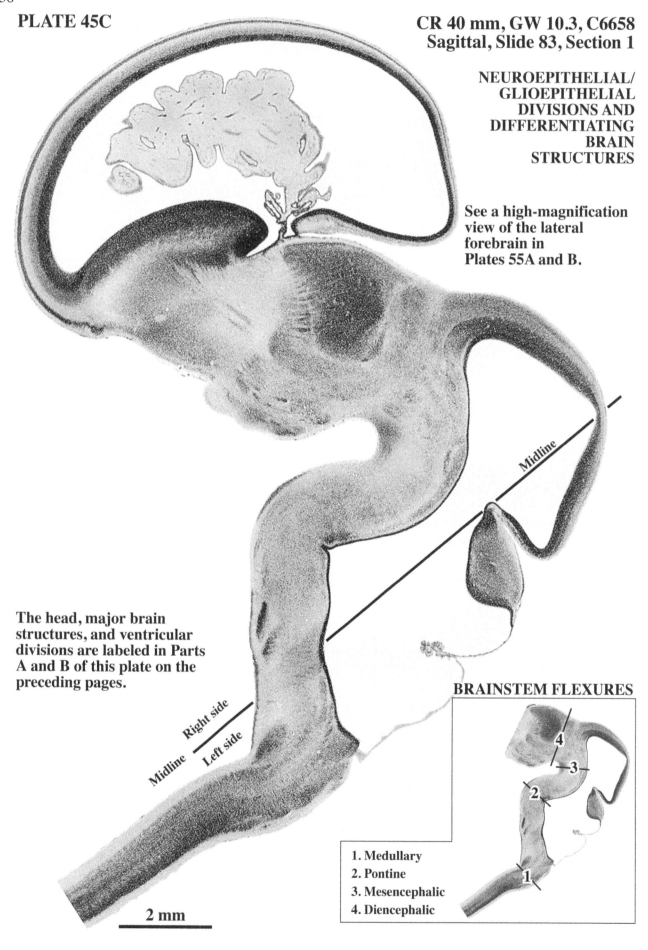

Midline

The head, major brain
structures, and ventricular
divisions are labeled in Parts
A and B of this plate on the
preceding pages.

Right side
Midline   Left side

BRAINSTEM FLEXURES

4
3
2
1

1. Medullary
2. Pontine
3. Mesencephalic
4. Diencephalic

2 mm

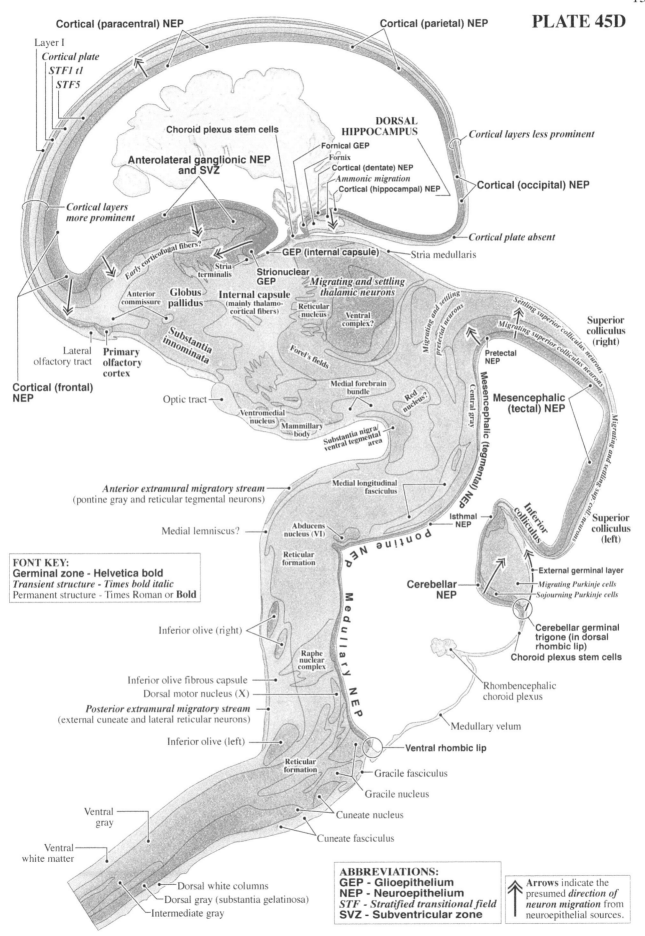

PLATE 45D

Cortical (paracentral) NEP

Cortical (parietal) NEP

Layer I
Cortical plate
STF1 t1
STF5

Choroid plexus stem cells

DORSAL HIPPOCAMPUS

Fornical GEP
Fornix
Cortical (dentate) NEP
Ammonic migration
Cortical (hippocampal) NEP

Cortical layers less prominent

Anterolateral ganglionic NEP and SVZ

Cortical (occipital) NEP

Cortical layers more prominent

Cortical plate absent

Early corticofugal fibers?

GEP (internal capsule)

Stria medullaris

Stria terminalis

Strionuclear GEP

Migrating and settling thalamic neurons

Superior colliculus (right)

Anterior commissure

Globus pallidus

Internal capsule (mainly thalamo-cortical fibers)

Reticular nucleus

Ventral complex?

Migrating and settling pretectal neurons

Settling superior colliculus neurons

Migrating superior colliculus neurons

Lateral olfactory tract

Primary olfactory cortex

Substantia innominata

Forel's fields

Mesencephalic (tegmental) NEP

Pretectal NEP

Mesencephalic (tectal) NEP

Cortical (frontal) NEP

Optic tract

Medial forebrain bundle

Red nucleus?

Central gray

Migrating and settling inf. coll. neurons

Ventromedial nucleus

Mammillary body

Substantia nigra/ ventral tegmental area

Anterior extramural migratory stream
(pontine gray and reticular tegmental neurons)

Medial longitudinal fasciculus

Inferior colliculus

Isthmal NEP

Superior colliculus (left)

Medial lemniscus?

Abducens nucleus (VI)

Pontine NEP

FONT KEY:
Germinal zone - Helvetica bold
Transient structure - Times bold italic
Permanent structure - Times Roman or **Bold**

Reticular formation

External germinal layer

Migrating Purkinje cells
Sojourning Purkinje cells

Cerebellar NEP

Cerebellar germinal trigone (in dorsal rhombic lip)

Choroid plexus stem cells

Inferior olive (right)

Medullary NEP

Raphe nuclear complex

Rhombencephalic choroid plexus

Inferior olive fibrous capsule

Dorsal motor nucleus (X)

Medullary velum

Posterior extramural migratory stream
(external cuneate and lateral reticular neurons)

Inferior olive (left)

Ventral rhombic lip

Reticular formation

Gracile fasciculus

Gracile nucleus

Ventral gray

Cuneate nucleus

Ventral white matter

Cuneate fasciculus

Dorsal white columns
Dorsal gray (substantia gelatinosa)
Intermediate gray

ABBREVIATIONS:
GEP - Glioepithelium
NEP - Neuroepithelium
STF - Stratified transitional field
SVZ - Subventricular zone

Arrows indicate the presumed *direction of neuron migration* from neuroepithelial sources.

PLATE 46A
CR 40 mm, GW 10.3, C6658
Sagittal, Slide 95, Section 1

HEAD STRUCTURES,
MAJOR BRAIN REGIONS,
AND VENTRICULAR DIVISIONS

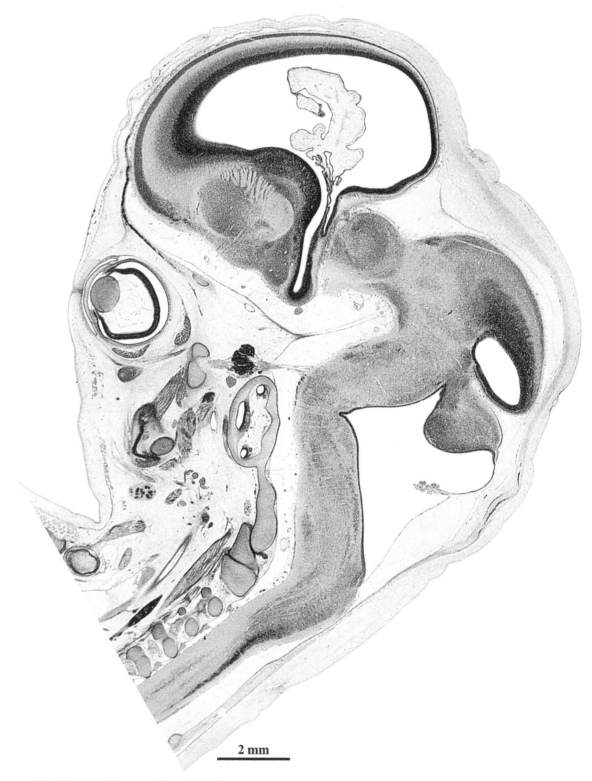

2 mm

Neuroepithelial divisions, glioepithelial divisions, and differentiating
structures are labeled in Parts C and D of this plate on the following pages.

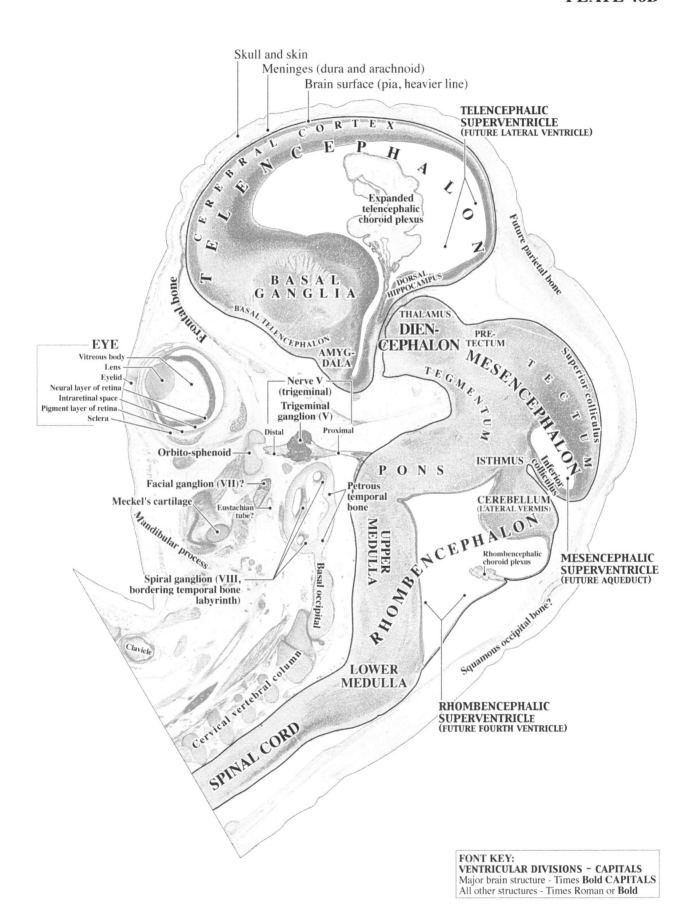

Skull and skin
Meninges (dura and arachnoid)
Brain surface (pia, heavier line)

**TELENCEPHALIC
SUPERVENTRICLE**
(FUTURE LATERAL VENTRICLE)

C E R E B R A L   C O R T E X

T E L E N C E P H A L O N

Future parietal bone

**Expanded
telencephalic
choroid plexus**

DORSAL
HIPPOCAMPUS

THALAMUS
**DIEN-
CEPHALON**

PRE-
TECTUM

BASAL
GANGLIA

BASAL TELENCEPHALON

**AMYG-
DALA**

MESENCEPHALON

T E G M E N T U M

T E C T U M

Superior colliculus

Frontal bone

EYE
Vitreous body
Lens
Eyelid
Neural layer of retina
Intraretinal space
Pigment layer of retina
Sclera

**Nerve V
(trigeminal)**

**Trigeminal
ganglion (V)**

Distal   Proximal

Orbito-sphenoid

ISTHMUS

Inferior
colliculus

P O N S

CEREBELLUM
(LATERAL VERMIS)

Facial ganglion (VII)?

Meckel's cartilage

Eustachian
tube?

**Petrous
temporal
bone**

UPPER
MEDULLA

**Rhombencephalic
choroid plexus**

**MESENCEPHALIC
SUPERVENTRICLE**
(FUTURE AQUEDUCT)

Mandibular process

Basal occipital

**Spiral ganglion (VIII,
bordering temporal bone
labyrinth)**

R H O M B E N C E P H A L O N

Squamous occipital bone?

Clavicle

LOWER
MEDULLA

Cervical vertebral column

SPINAL CORD

**RHOMBENCEPHALIC
SUPERVENTRICLE**
(FUTURE FOURTH VENTRICLE)

FONT KEY:
VENTRICULAR DIVISIONS – CAPITALS
Major brain structure - Times **Bold CAPITALS**
All other structures - Times Roman or **Bold**

PLATE 46C

**NEUROEPITHELIAL/
GLIOEPITHELIAL
DIVISIONS AND
DIFFERENTIATING
BRAIN STRUCTURES**

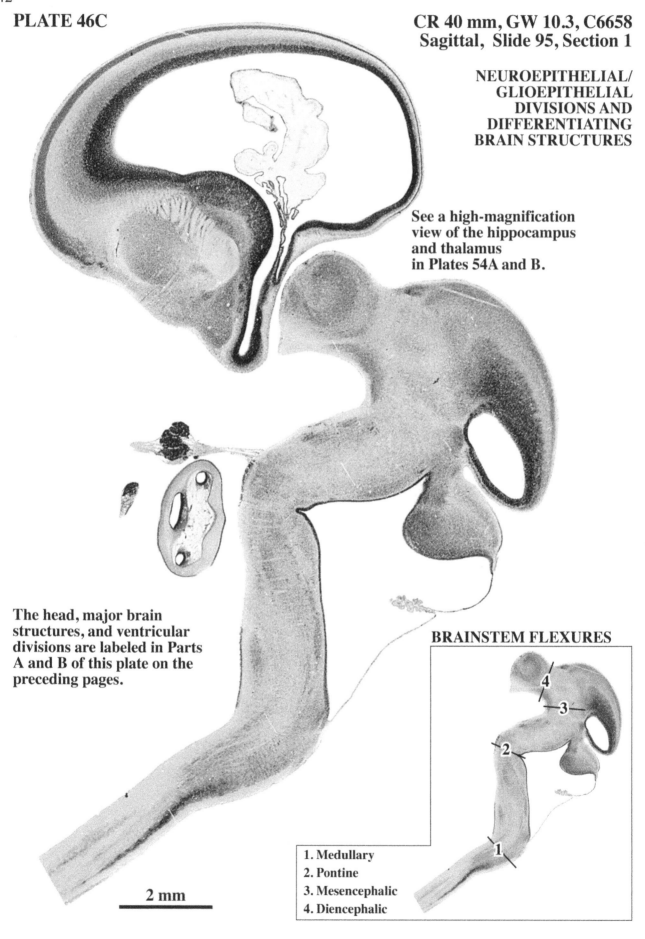

See a high-magnification
view of the hippocampus
and thalamus
in Plates 54A and B.

The head, major brain
structures, and ventricular
divisions are labeled in Parts
A and B of this plate on the
preceding pages.

BRAINSTEM FLEXURES

1. Medullary
2. Pontine
3. Mesencephalic
4. Diencephalic

2 mm

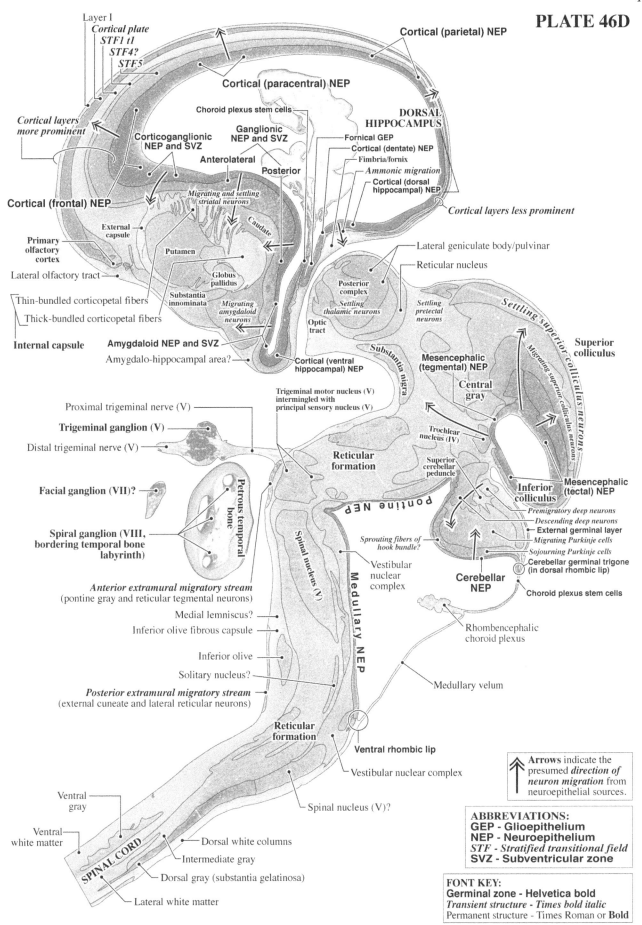

Layer I
Cortical plate
STF1 t1
STF4?
STF5

Cortical (parietal) NEP

Cortical (paracentral) NEP

Choroid plexus stem cells

Ganglionic NEP and SVZ

DORSAL HIPPOCAMPUS

Fornical GEP

Cortical (dentate) NEP

Fimbria/fornix

Ammonic migration

Cortical (dorsal hippocampal) NEP

Cortical layers more prominent

Corticoganglionic NEP and SVZ

Anterolateral

Posterior

Migrating and settling striatal neurons

Cortical (frontal) NEP

Cortical layers less prominent

External capsule

Caudate

Lateral geniculate body/pulvinar

Reticular nucleus

Primary olfactory cortex

Putamen

Posterior complex

Settling thalamic neurons

Settling pretectal neurons

Settling superior colliculus neurons

Lateral olfactory tract

Globus pallidus

Substantia innominata

Migrating amygdaloid neurons

Superior colliculus

Thin-bundled corticopetal fibers

Thick-bundled corticopetal fibers

Optic tract

Migrating superior colliculus neurons

Internal capsule

Amygdaloid NEP and SVZ

Amygdalo-hippocampal area?

Cortical (ventral hippocampal) NEP

Substantia nigra

Mesencephalic (tegmental) NEP

Central gray

Trigeminal motor nucleus (V) intermingled with principal sensory nucleus (V)

Proximal trigeminal nerve (V)

Trochlear nucleus (IV)

Mesencephalic (tectal) NEP

Trigeminal ganglion (V)

Superior cerebellar peduncle

Inferior colliculus

Distal trigeminal nerve (V)

Reticular formation

Premigratory deep neurons

Descending deep neurons

Facial ganglion (VII)?

Pontine NEP

External germinal layer

Migrating Purkinje cells

Spiral ganglion (VIII, bordering temporal bone labyrinth)

Petrous temporal bone

Sprouting fibers of hook bundle?

Sojourning Purkinje cells

Cerebellar germinal trigone (in dorsal rhombic lip)

Cerebellar NEP

Anterior extramural migratory stream
(pontine gray and reticular tegmental neurons)

Spinal nucleus (V)

Vestibular nuclear complex

Choroid plexus stem cells

Medial lemniscus?

Inferior olive fibrous capsule

Medullary NEP

Rhombencephalic choroid plexus

Inferior olive

Solitary nucleus?

Posterior extramural migratory stream
(external cuneate and lateral reticular neurons)

Reticular formation

Medullary velum

Ventral rhombic lip

Vestibular nuclear complex

Ventral gray

Spinal nucleus (V)?

Ventral white matter

Dorsal white columns

Intermediate gray

SPINAL CORD

Dorsal gray (substantia gelatinosa)

Lateral white matter

Arrows indicate the presumed *direction of neuron migration* from neuroepithelial sources.

ABBREVIATIONS:
GEP - Glioepithelium
NEP - Neuroepithelium
STF - *Stratified transitional field*
SVZ - Subventricular zone

FONT KEY:
Germinal zone - Helvetica bold
Transient structure - Times bold italic
Permanent structure - Times Roman or **Bold**

PLATE 47A
CR 40 mm, GW 10.3, C6658
Sagittal, Slide 99, Section 1

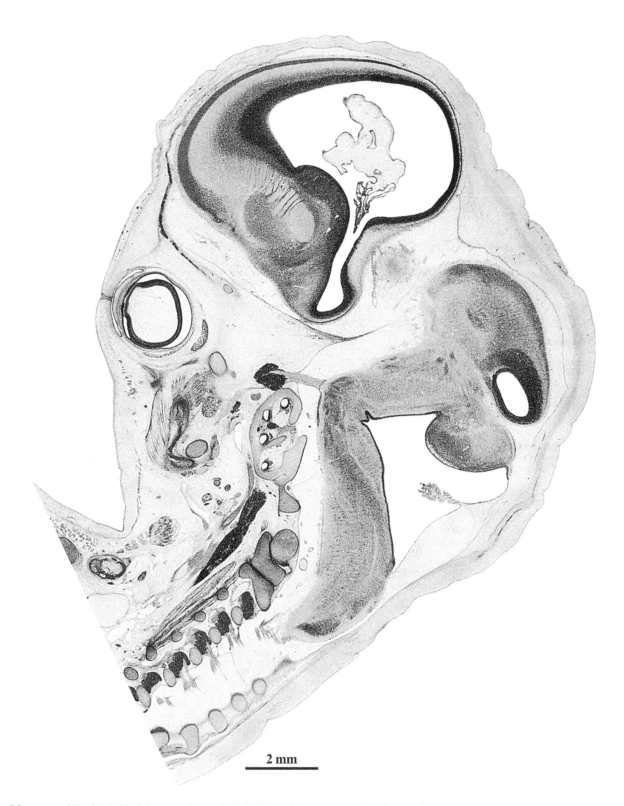

2 mm

Neuroepithelial divisions, glioepithelial divisions, and differentiating
structures are labeled in Parts C and D of this plate on the following pages.

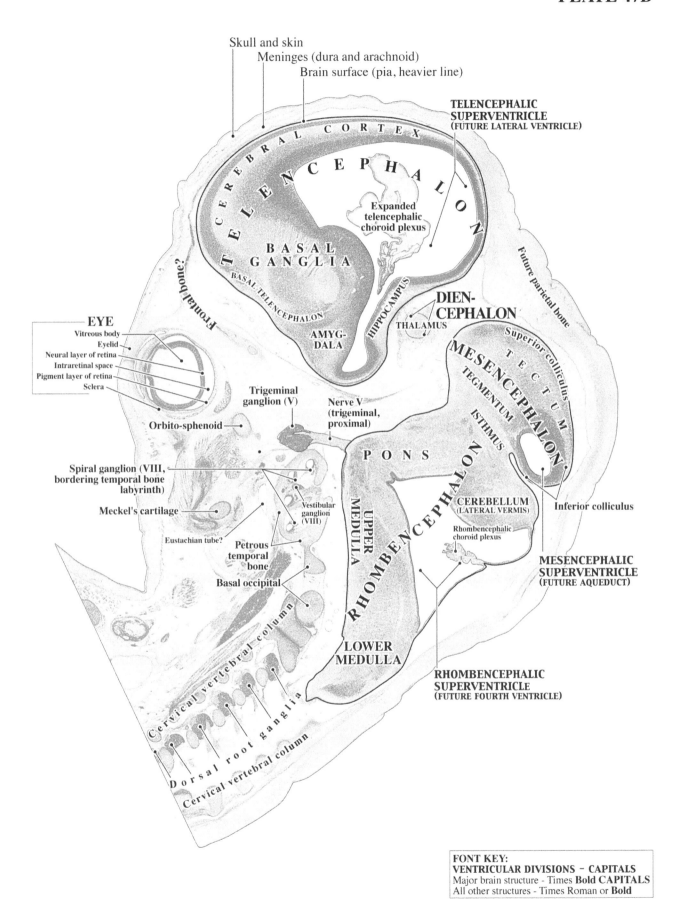

Skull and skin
Meninges (dura and arachnoid)
Brain surface (pia, heavier line)

**TELENCEPHALIC
SUPERVENTRICLE
(FUTURE LATERAL VENTRICLE)**

C E R E B R A L   C O R T E X

T E L E N C E P H A L O N

Expanded
telencephalic
choroid plexus

Future parietal bone

**BASAL
GANGLIA**

BASAL TELENCEPHALON

Frontal bone?

HIPPOCAMPUS

**DIEN-
CEPHALON**

THALAMUS

**AMYG-
DALA**

Superior colliculus

T E C T U M

MESENCEPHALON

TEGMENTUM

ISTHMUS

EYE
Vitreous body
Eyelid
Neural layer of retina
Intraretinal space
Pigment layer of retina
Sclera

**Trigeminal
ganglion (V)**

**Nerve V
(trigeminal,
proximal)**

P O N S

Orbito-sphenoid

**Spiral ganglion (VIII,
bordering temporal bone
labyrinth)**

UPPER
MEDULLA

R H O M B E N C E P H A L O N

**CEREBELLUM
(LATERAL VERMIS)**

Inferior colliculus

Meckel's cartilage

**Vestibular
ganglion
(VIII)**

Rhombencephalic
choroid plexus

Eustachian tube?

**Petrous
temporal
bone**

**MESENCEPHALIC
SUPERVENTRICLE
(FUTURE AQUEDUCT)**

Basal occipital

**LOWER
MEDULLA**

**RHOMBENCEPHALIC
SUPERVENTRICLE
(FUTURE FOURTH VENTRICLE)**

Cervical vertebral column

D o r s a l   r o o t   g a n g l i a

Cervical vertebral column

FONT KEY:
VENTRICULAR DIVISIONS – CAPITALS
Major brain structure - Times **Bold CAPITALS**
All other structures - Times Roman or **Bold**

PLATE 47C

**NEUROEPITHELIAL/GLIOEPITHELIAL
DIVISIONS AND DIFFERENTIATING
BRAIN STRUCTURES**

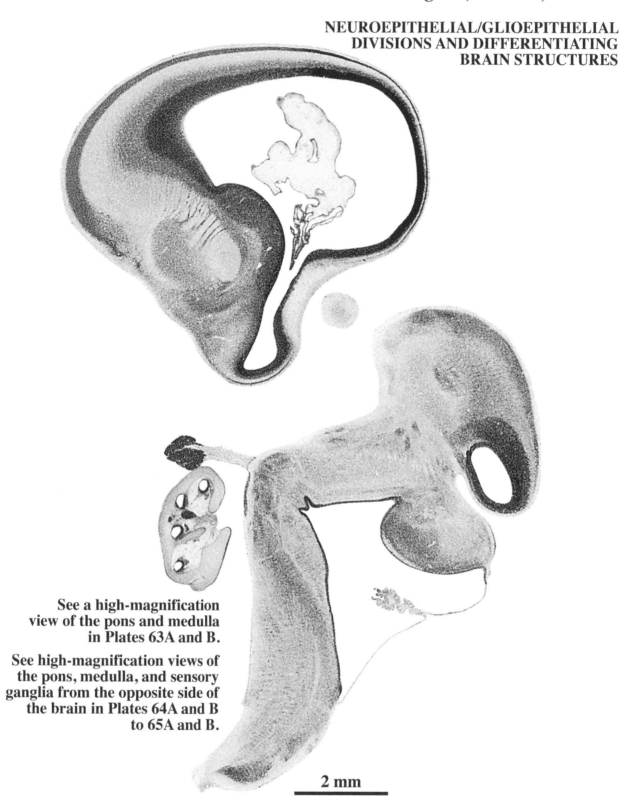

See a high-magnification
view of the pons and medulla
in Plates 63A and B.

See high-magnification views of
the pons, medulla, and sensory
ganglia from the opposite side of
the brain in Plates 64A and B
to 65A and B.

2 mm

The head, major brain structures, and ventricular divisions are
labeled in Parts A and B of this plate on the preceding pages.

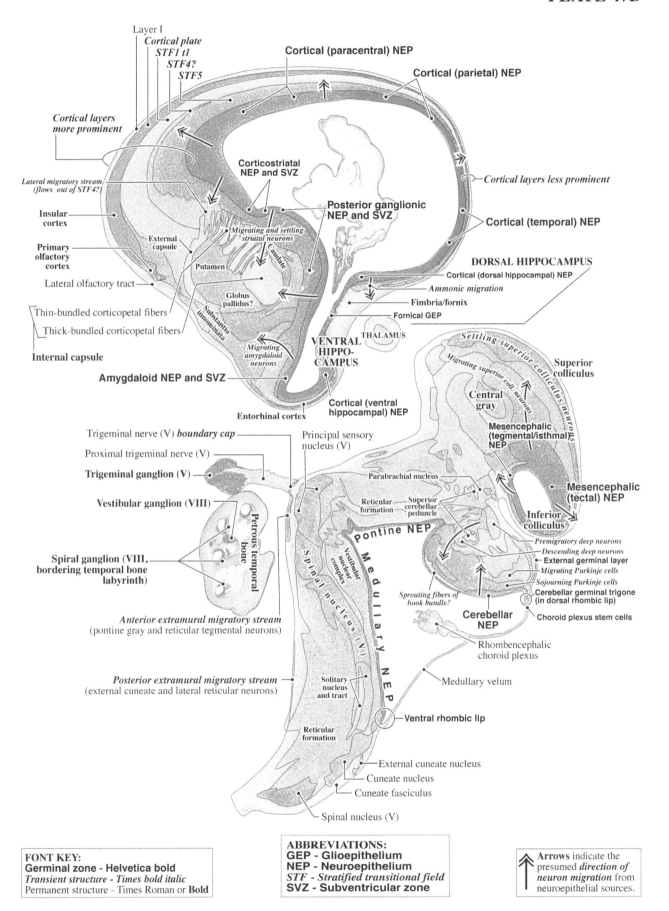

Layer I
Cortical plate
STF1 t1
STF4?
STF5

Cortical (paracentral) NEP

Cortical (parietal) NEP

Cortical layers more prominent

Cortical layers less prominent

Lateral migratory stream (flows out of STF4?)

Insular cortex

Corticostriatal NEP and SVZ

Posterior ganglionic NEP and SVZ

Cortical (temporal) NEP

Primary olfactory cortex

External capsule

Migrating and settling striatal neurons

DORSAL HIPPOCAMPUS

Cortical (dorsal hippocampal) NEP

Lateral olfactory tract

Putamen

Caudate

Ammonic migration

Thin-bundled corticopetal fibers

Globus pallidus?

Fimbria/fornix

Fornical GEP

Thick-bundled corticopetal fibers

Substantia innominata

THALAMUS

Settling superior colliculus

Superior colliculus

Internal capsule

Migrating amygdaloid neurons

VENTRAL HIPPO-CAMPUS

Migrating superior coll. neurons

Central gray

Amygdaloid NEP and SVZ

Mesencephalic (tegmental/isthmal) NEP

Entorhinal cortex

Cortical (ventral hippocampal) NEP

Trigeminal nerve (V) *boundary cap*

Principal sensory nucleus (V)

Proximal trigeminal nerve (V)

Mesencephalic (tectal) NEP

Trigeminal ganglion (V)

Parabrachial nucleus

Vestibular ganglion (VIII)

Reticular formation

Superior cerebellar peduncle

Inferior colliculus

Petrous temporal bone

Pontine NEP

Premigratory deep neurons
Descending deep neurons
External germinal layer
Migrating Purkinje cells

Spiral ganglion (VIII, bordering temporal bone labyrinth)

Vestibular nuclear complex

Spinal nucleus (V)

Medullary NEP

Sojourning Purkinje cells
Cerebellar germinal trigone (in dorsal rhombic lip)

Sprouting fibers of hook bundle?

Cerebellar NEP

Choroid plexus stem cells

Anterior extramural migratory stream (pontine gray and reticular tegmental neurons)

Rhombencephalic choroid plexus

Medullary velum

Posterior extramural migratory stream (external cuneate and lateral reticular neurons)

Solitary nucleus and tract

Ventral rhombic lip

Reticular formation

External cuneate nucleus

Cuneate nucleus

Cuneate fasciculus

Spinal nucleus (V)

ABBREVIATIONS:
GEP - Glioepithelium
NEP - Neuroepithelium
STF - *Stratified transitional field*
SVZ - Subventricular zone

Arrows indicate the presumed *direction of neuron migration* from neuroepithelial sources.

PLATE 48A
CR 40 mm, GW 10.3, C6658
Sagittal, Slide 103, Section 1

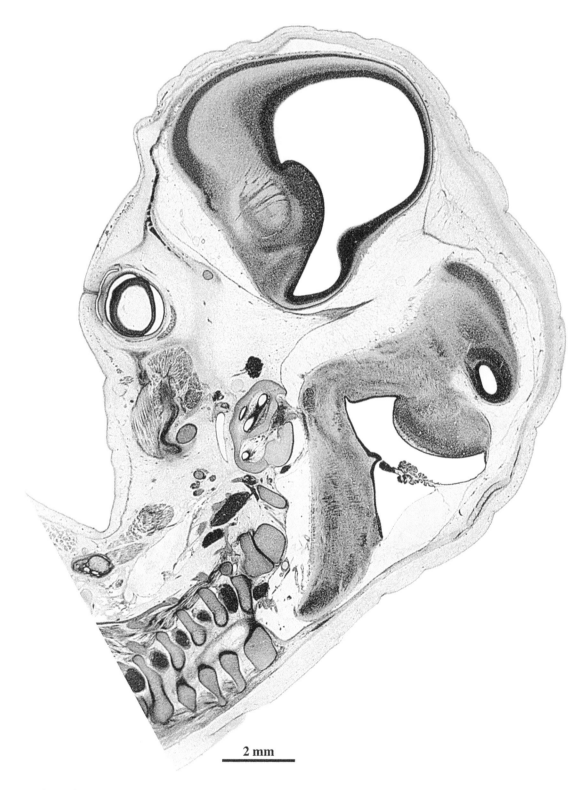

2 mm

**Neuroepithelial divisions, glioepithelial divisions, and differentiating
structures are labeled in Parts C and D of this plate on the following pages.**

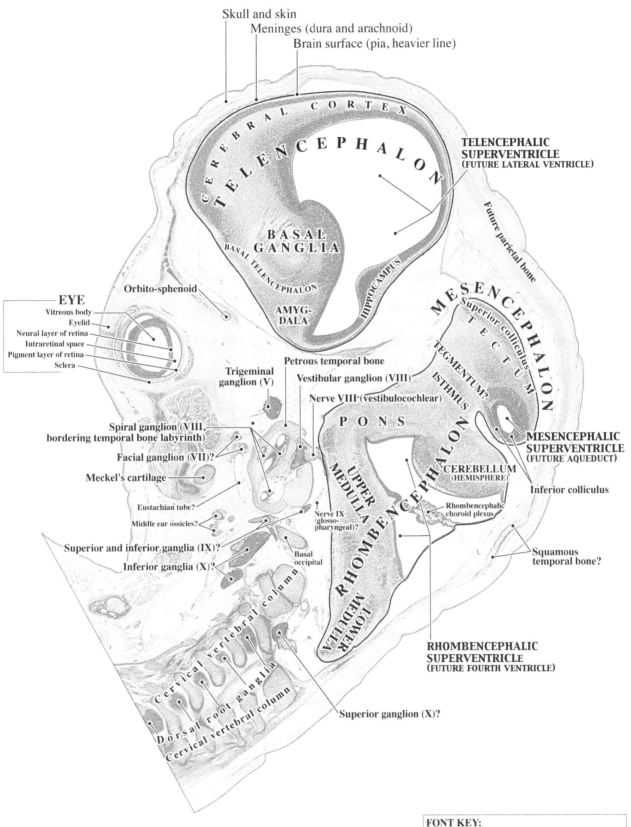

Skull and skin
Meninges (dura and arachnoid)
Brain surface (pia, heavier line)

C E R E B R A L   C O R T E X

TELENCEPHALON

**TELENCEPHALIC
SUPERVENTRICLE
(FUTURE LATERAL VENTRICLE)**

Future parietal bone

**BASAL
GANGLIA**

BASAL TELENCEPHALON

HIPPOCAMPUS

**AMYG-
DALA**

MESENCEPHALON

Superior colliculus

T E C T U M

EYE
Orbito-sphenoid
Vitreous body
Eyelid
Neural layer of retina
Intraretinal space
Pigment layer of retina
Sclera

TEGMENTUM?

ISTHMUS

Petrous temporal bone

Trigeminal
ganglion (V)

Vestibular ganglion (VIII)

Nerve VIII (vestibulocochlear)

P O N S

**MESENCEPHALIC
SUPERVENTRICLE
(FUTURE AQUEDUCT)**

Spiral ganglion (VIII,
bordering temporal bone labyrinth)

Facial ganglion (VII)?

Meckel's cartilage

**CEREBELLUM
(HEMISPHERE)**

Inferior colliculus

UPPER
MEDULLA

RHOMBENCEPHALON

Rhombencephalic
choroid plexus

Eustachian tube?

Middle ear ossicles?

Nerve IX
(glosso-
pharyngeal)?

Squamous
temporal bone?

Superior and inferior ganglia (IX)?

Inferior ganglia (X)?

Basal
occipital

LOWER MEDULLA

**RHOMBENCEPHALIC
SUPERVENTRICLE
(FUTURE FOURTH VENTRICLE)**

Cervical vertebral column

Dorsal root ganglia

Cervical vertebral column

Superior ganglion (X)?

FONT KEY:
VENTRICULAR DIVISIONS – CAPITALS
Major brain structure - Times **Bold CAPITALS**
All other structures - Times Roman or **Bold**

PLATE 48C
CR 40 mm, GW 10.3, C6658
Sagittal, Slide 103, Section 1

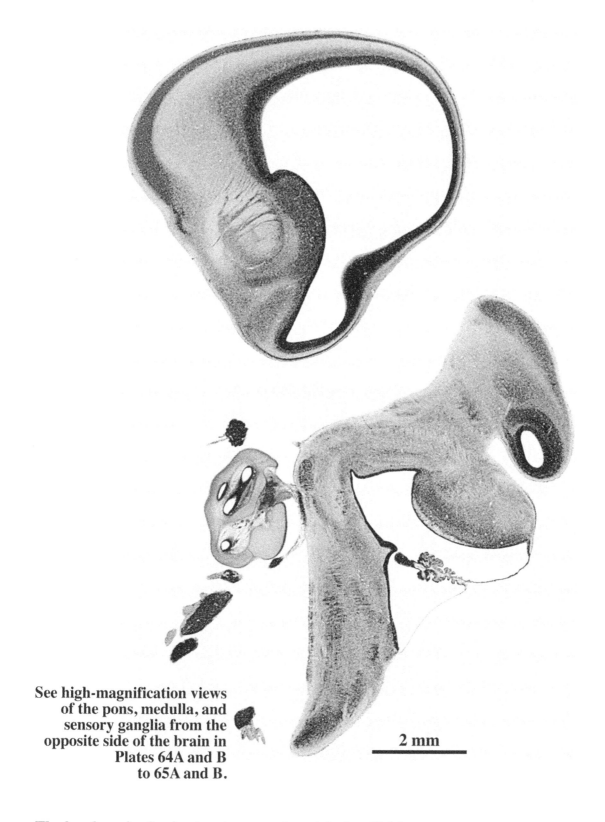

See high-magnification views
of the pons, medulla, and
sensory ganglia from the
opposite side of the brain in
Plates 64A and B
to 65A and B.

2 mm

The head, major brain structures, and ventricular divisions are
labeled in Parts A and B of this plate on the preceding pages.

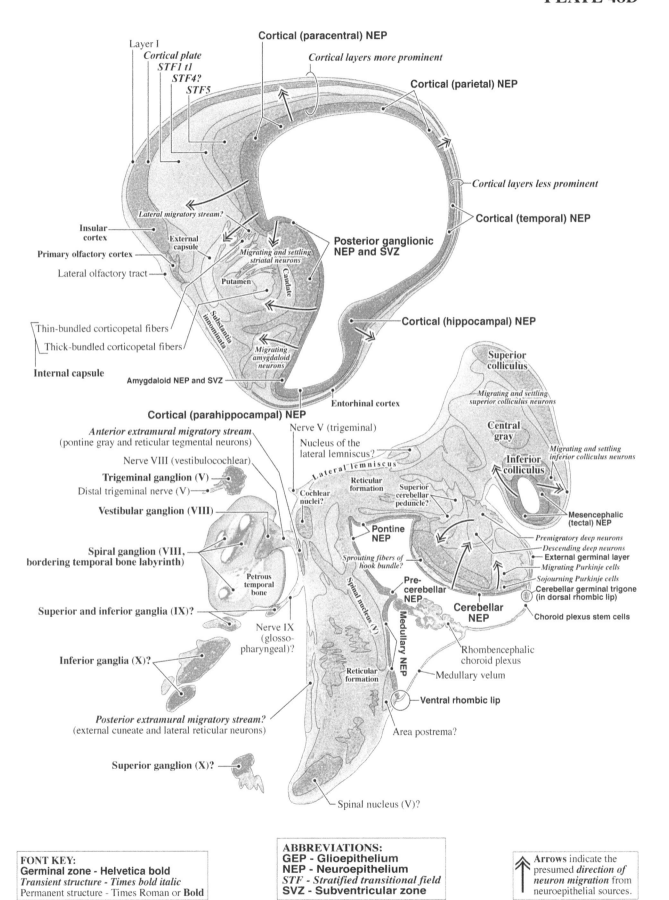

Layer I
Cortical plate
STF1 t1
STF4?
STF5

Cortical (paracentral) NEP

Cortical layers more prominent

Cortical (parietal) NEP

Cortical layers less prominent

Cortical (temporal) NEP

Lateral migratory stream?

Insular cortex

External capsule

Primary olfactory cortex

Lateral olfactory tract

Migrating and settling striatal neurons

Putamen

Caudate

Posterior ganglionic NEP and SVZ

Cortical (hippocampal) NEP

Thin-bundled corticopetal fibers

Thick-bundled corticopetal fibers

Substantia innominata

Migrating amygdaloid neurons

Internal capsule

Amygdaloid NEP and SVZ

Entorhinal cortex

Cortical (parahippocampal) NEP

Superior colliculus

Migrating and settling superior colliculus neurons

Central gray

Inferior colliculus

Migrating and settling inferior colliculus neurons

Anterior extramural migratory stream
(pontine gray and reticular tegmental neurons)

Nerve V (trigeminal)

Nucleus of the lateral lemniscus?

Nerve VIII (vestibulocochlear)

Lateral lemniscus

Reticular formation

Superior cerebellar peduncle?

Mesencephalic (tectal) NEP

Trigeminal ganglion (V)

Distal trigeminal nerve (V)

Cochlear nuclei?

Premigratory deep neurons
Descending deep neurons
External germinal layer
Migrating Purkinje cells

Vestibular ganglion (VIII)

Pontine NEP

Sojourning Purkinje cells
Cerebellar germinal trigone (in dorsal rhombic lip)

Spiral ganglion (VIII, bordering temporal bone labyrinth)

Sprouting fibers of hook bundle?

Pre-cerebellar NEP

Cerebellar NEP

Choroid plexus stem cells

Petrous temporal bone

Spinal nucleus (V)

Medullary NEP

Superior and inferior ganglia (IX)?

Nerve IX (glosso-pharyngeal)?

Rhombencephalic choroid plexus

Medullary velum

Inferior ganglia (X)?

Reticular formation

Ventral rhombic lip

Posterior extramural migratory stream?
(external cuneate and lateral reticular neurons)

Area postrema?

Superior ganglion (X)?

Spinal nucleus (V)?

FONT KEY:
Germinal zone - Helvetica bold
Transient structure - Times bold italic
Permanent structure - Times Roman or **Bold**

ABBREVIATIONS:
GEP - Glioepithelium
NEP - Neuroepithelium
STF - Stratified transitional field
SVZ - Subventricular zone

Arrows indicate the presumed *direction of neuron migration* from neuroepithelial sources.

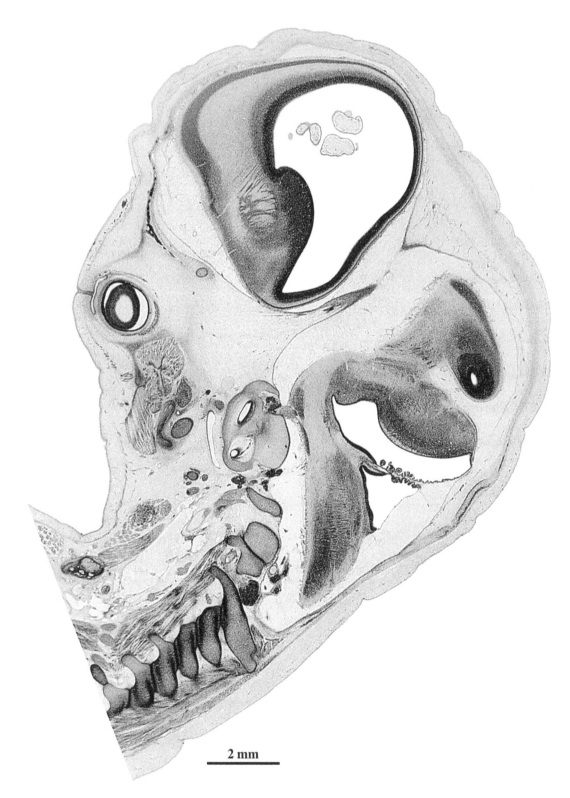

2 mm

Neuroepithelial divisions, glioepithelial divisions, and differentiating
structures are labeled in Parts C and D of this plate on the following pages.

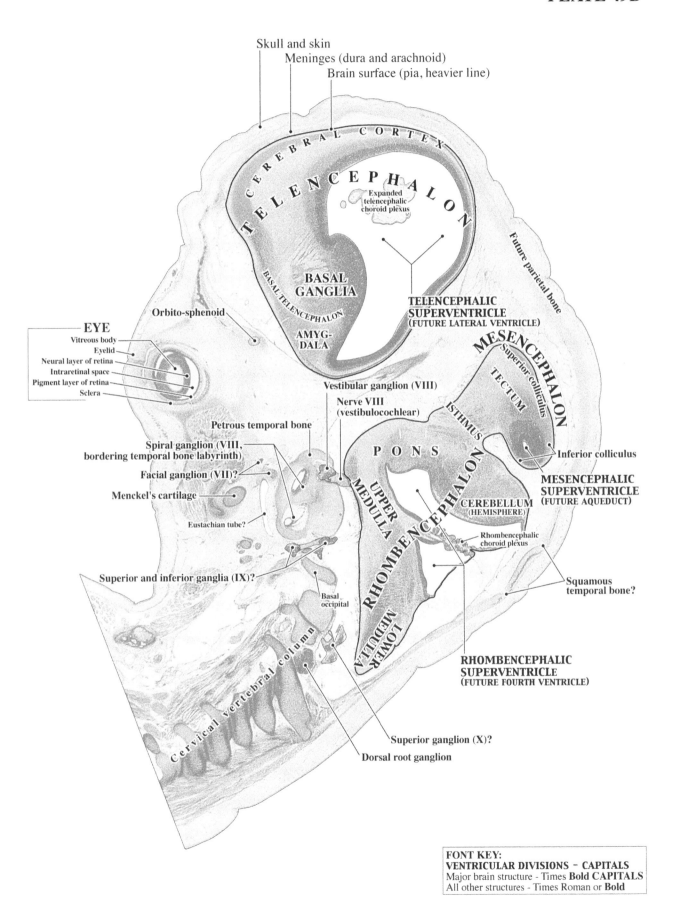

Skull and skin
Meninges (dura and arachnoid)
Brain surface (pia, heavier line)

C E R E B R A L   C O R T E X

T E L E N C E P H A L O N

Expanded
telencephalic
choroid plexus

BASAL TELENCEPHALON

Future parietal bone

**BASAL
GANGLIA**

**TELENCEPHALIC
SUPERVENTRICLE**
(FUTURE LATERAL VENTRICLE)

MESENCEPHALON

Superior colliculus

EYE

Orbito-sphenoid

**AMYG-
DALA**

Vitreous body
Eyelid
Neural layer of retina
Intraretinal space
Pigment layer of retina
Sclera

Vestibular ganglion (VIII)

Nerve VIII
(vestibulocochlear)

TECTUM

ISTHMUS

P O N S

Inferior colliculus

Petrous temporal bone

**Spiral ganglion (VIII,
bordering temporal bone labyrinth)**

Facial ganglion (VII)?

Menckel's cartilage

Eustachian tube?

UPPER
MEDULLA

R H O M B E N C E P H A L O N

CEREBELLUM
(HEMISPHERE)

**MESENCEPHALIC
SUPERVENTRICLE**
(FUTURE AQUEDUCT)

Rhombencephalic
choroid plexus

Squamous
temporal bone?

Superior and inferior ganglia (IX)?

Basal
occipital

LOWER
MEDULLA

**RHOMBENCEPHALIC
SUPERVENTRICLE**
(FUTURE FOURTH VENTRICLE)

Cervical vertebral column

Superior ganglion (X)?

Dorsal root ganglion

PLATE 49C
CR 40 mm, GW 10.3, C6658
Sagittal, Slide 105, Section 2

NEUROEPITHELIAL/GLIOEPITHELIAL
DIVISIONS AND DIFFERENTIATING
BRAIN STRUCTURES

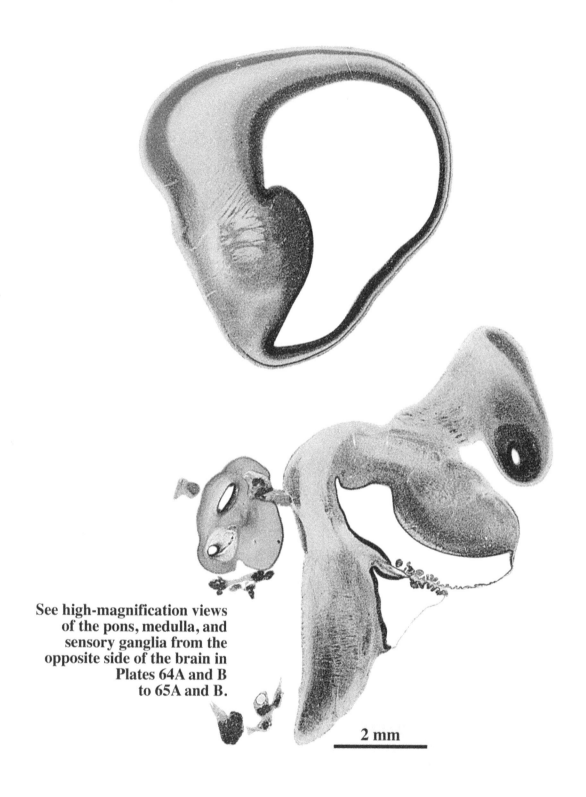

See high-magnification views
of the pons, medulla, and
sensory ganglia from the
opposite side of the brain in
Plates 64A and B
to 65A and B.

2 mm

The head, major brain structures, and ventricular divisions are
labeled in Parts A and B of this plate on the preceding pages.

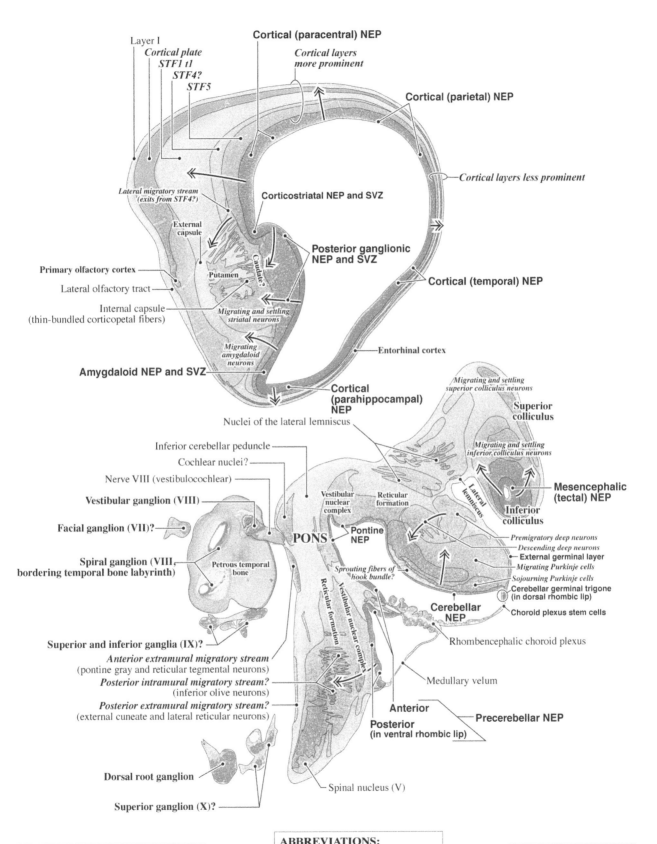

Layer I
Cortical plate
STF1 t1
STF4?
STF5

Cortical (paracentral) NEP

Cortical layers more prominent

Cortical (parietal) NEP

Cortical layers less prominent

Lateral migratory stream (exits from STF4?)

Corticostriatal NEP and SVZ

External capsule

Posterior ganglionic NEP and SVZ

Cortical (temporal) NEP

Primary olfactory cortex

Caudate?

Putamen

Lateral olfactory tract

Internal capsule
(thin-bundled corticopetal fibers)

Migrating and settling striatal neurons

Migrating amygdaloid neurons

Entorhinal cortex

Amygdaloid NEP and SVZ

Cortical (parahippocampal) NEP

Migrating and settling superior colliculus neurons

Superior colliculus

Nuclei of the lateral lemniscus

Migrating and settling inferior colliculus neurons

Inferior cerebellar peduncle

Cochlear nuclei?

Nerve VIII (vestibulocochlear)

Vestibular nuclear complex

Reticular formation

Lateral lemniscus

Mesencephalic (tectal) NEP

Inferior colliculus

Vestibular ganglion (VIII)

Facial ganglion (VII)?

PONS

Pontine NEP

Premigratory deep neurons
Descending deep neurons
External germinal layer
Migrating Purkinje cells
Sojourning Purkinje cells
Cerebellar germinal trigone (in dorsal rhombic lip)

Spiral ganglion (VIII, bordering temporal bone labyrinth)

Petrous temporal bone

Sprouting fibers of hook bundle?

Vestibular nuclear complex

Cerebellar NEP

Choroid plexus stem cells

Rhombencephalic choroid plexus

Reticular formation

Superior and inferior ganglia (IX)?

Anterior extramural migratory stream
(pontine gray and reticular tegmental neurons)

Posterior intramural migratory stream?
(inferior olive neurons)

Posterior extramural migratory stream?
(external cuneate and lateral reticular neurons)

Medullary velum

Anterior

Posterior
(in ventral rhombic lip)

Precerebellar NEP

Dorsal root ganglion

Spinal nucleus (V)

Superior ganglion (X)?

0.05 mm

PLATE 50A
CR 40 mm
GW 10.3, C6658
Sagittal
Slide 71, Section 2

DORSAL
CORTEX

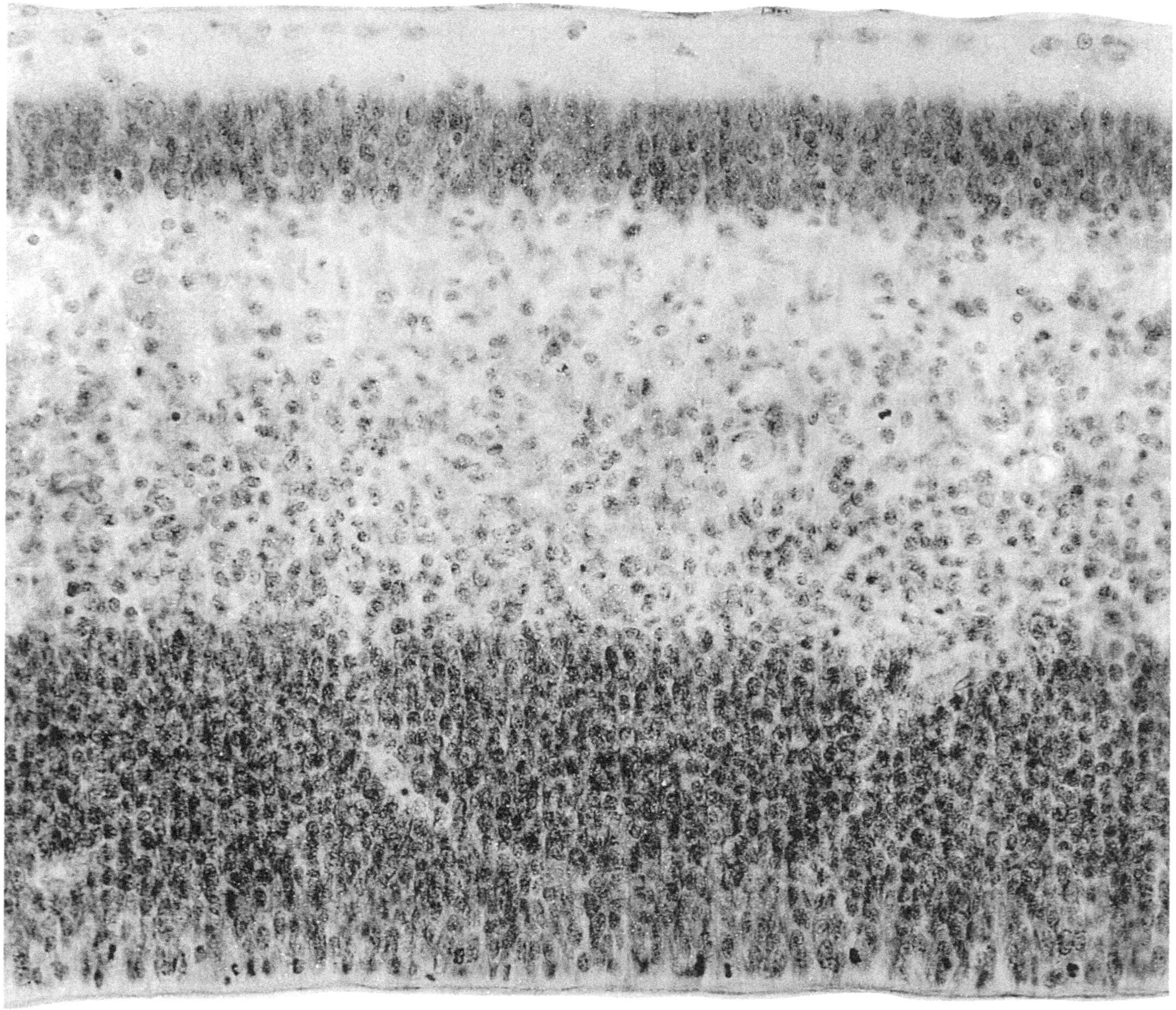

See the entire section in Plates 42A–D.

PLATE 50B

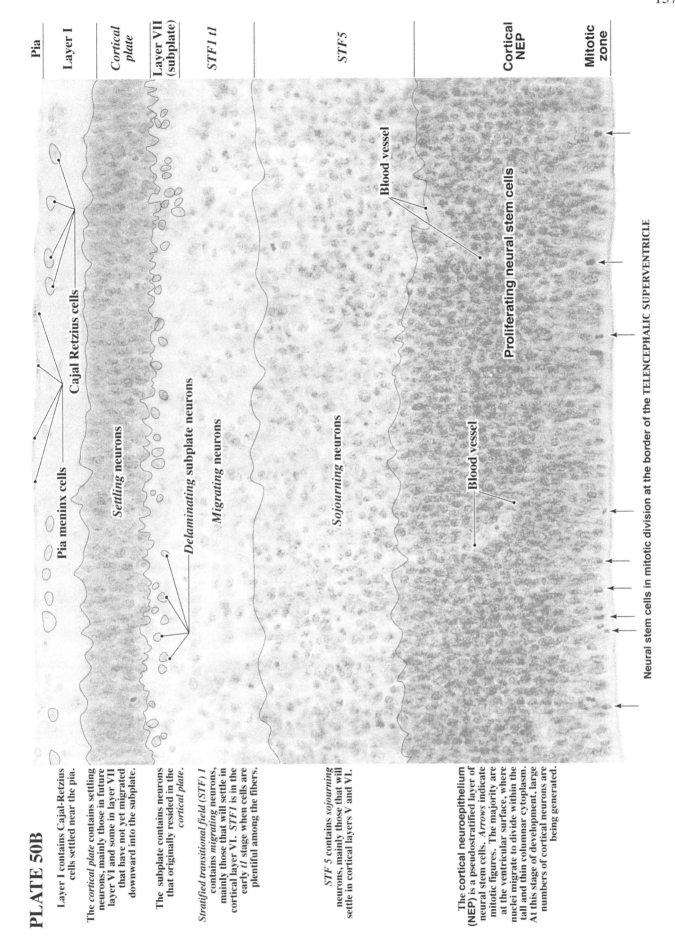

Pia

Layer I

Cortical plate

Layer VII (subplate)

STF1 t1

STF5

Cortical NEP

Mitotic zone

Cajal Retzius cells

Pia meninx cells

Settling neurons

Delaminating subplate neurons

Migrating neurons

Sojourning neurons

Blood vessel

Blood vessel

Proliferating neural stem cells

Neural stem cells in mitotic division at the border of the TELENCEPHALIC SUPERVENTRICLE

Layer I contains Cajal-Retzius cells settled near the pia.

The *cortical plate* contains settling neurons, mainly those in future layer VI and some in layer VII that have not yet migrated downward into the subplate.

The subplate contains neurons that originally resided in the *cortical plate.*

Stratified transitional field (STF) 1 contains *migrating* neurons, mainly those that will settle in cortical layer VI. *STF1* is in the early *t1* stage when cells are plentiful among the fibers.

STF 5 contains *sojourning* neurons, mainly those that will settle in cortical layers V and VI.

The cortical neuroepithelium (NEP) is a pseudostratified layer of neural stem cells. *Arrows* indicate mitotic figures. The majority are at the ventricular surface, where nuclei migrate to divide within the tall and thin columnar cytoplasm. At this stage of development, large numbers of cortical neurons are being generated.

158

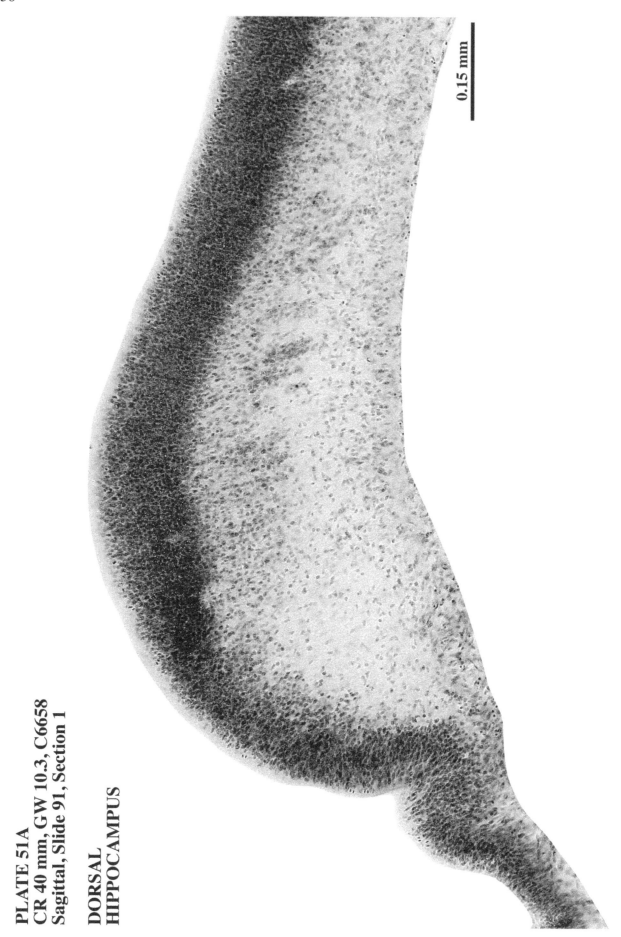

0.15 mm

PLATE 51A
CR 40 mm, GW 10.3, C6658
Sagittal, Slide 91, Section 1

DORSAL
HIPPOCAMPUS

See a low-magnification view of nearby sections in Plates 44A–D and 45A–D.

PLATE 51B

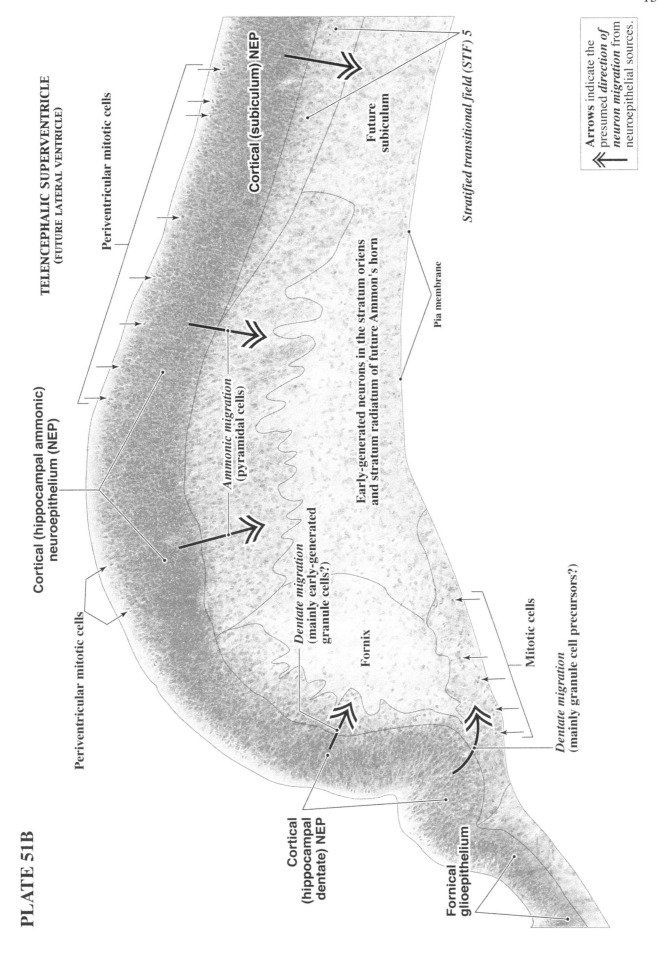

TELENCEPHALIC SUPERVENTRICLE
(FUTURE LATERAL VENTRICLE)

Periventricular mitotic cells

Cortical (subiculum) NEP

Stratified transitional field (STF) 5

Future subiculum

Pia membrane

Early-generated neurons in the stratum oriens
and stratum radiatum of future Ammon's horn

**Cortical (hippocampal ammonic)
neuroepithelium (NEP)**

Periventricular mitotic cells

*Ammonic migration
(pyramidal cells)*

*Dentate migration
(mainly early-generated
granule cells?)*

Fornix

Mitotic cells

*Dentate migration
(mainly granule cell precursors?)*

**Cortical
(hippocampal
dentate) NEP**

**Fornical
glioepithelium**

Arrows indicate the
presumed *direction of
neuron migration* from
neuroepithelial sources.

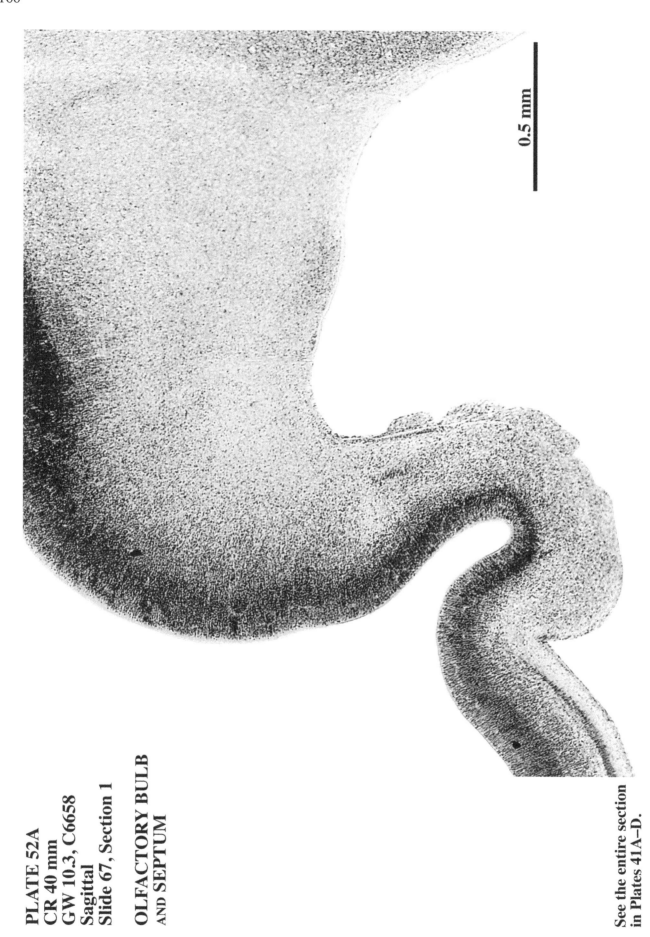

0.5 mm

PLATE 52A
CR 40 mm
GW 10.3, C6658
Sagittal
Slide 67, Section 1

OLFACTORY BULB
AND SEPTUM

See the entire section
in Plates 41A–D.

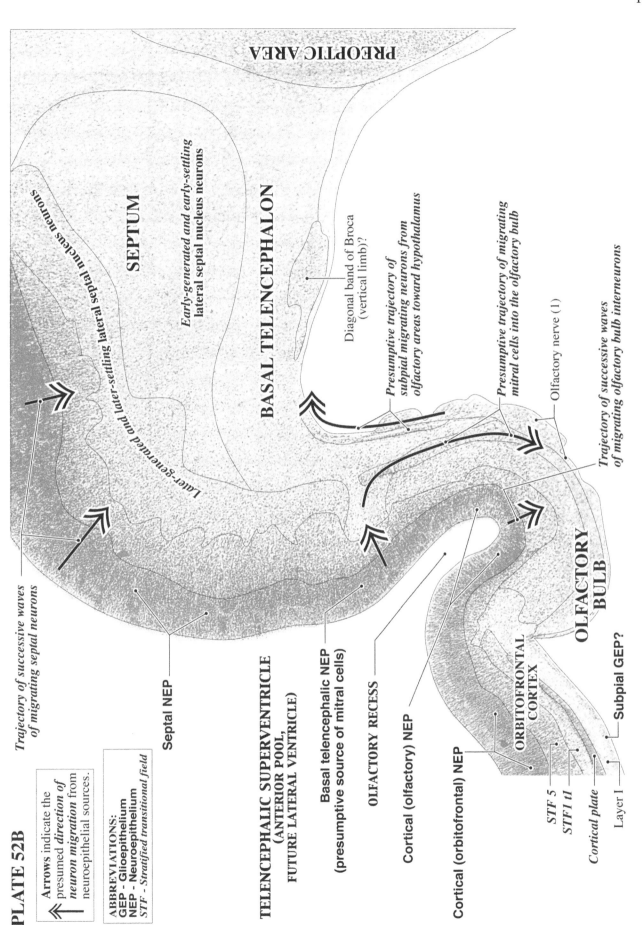

PLATE 52B

← Arrows indicate the *presumed direction of neuron migration* from neuroepithelial sources.

ABBREVIATIONS:
GEP - Glioepithelium
NEP - Neuroepithelium
STF - Stratified transitional field

PREOPTIC AREA

SEPTUM

Early-generated and early-settling lateral septal nucleus neurons

BASAL TELENCEPHALON

Diagonal band of Broca (vertical limb)?

Presumptive trajectory of subpial migrating neurons from olfactory areas toward hypothalamus

Presumptive trajectory of migrating mitral cells into the olfactory bulb

Olfactory nerve (1)

Trajectory of successive waves of migrating olfactory bulb interneurons

Trajectory of successive waves of migrating septal neurons

Later-generated and later-settling lateral septal nucleus neurons

OLFACTORY BULB

Trajectory of successive waves of migrating septal neurons

Septal NEP

TELENCEPHALIC SUPERVENTRICLE
(ANTERIOR POOL,
FUTURE LATERAL VENTRICLE)

Basal telencephalic NEP
(presumptive source of mitral cells)

OLFACTORY RECESS

Cortical (olfactory) NEP

Cortical (orbitofrontal) NEP

ORBITOFRONTAL CORTEX

STF 5
STF1 t1
Cortical plate
Layer 1

Subpial GEP?

162

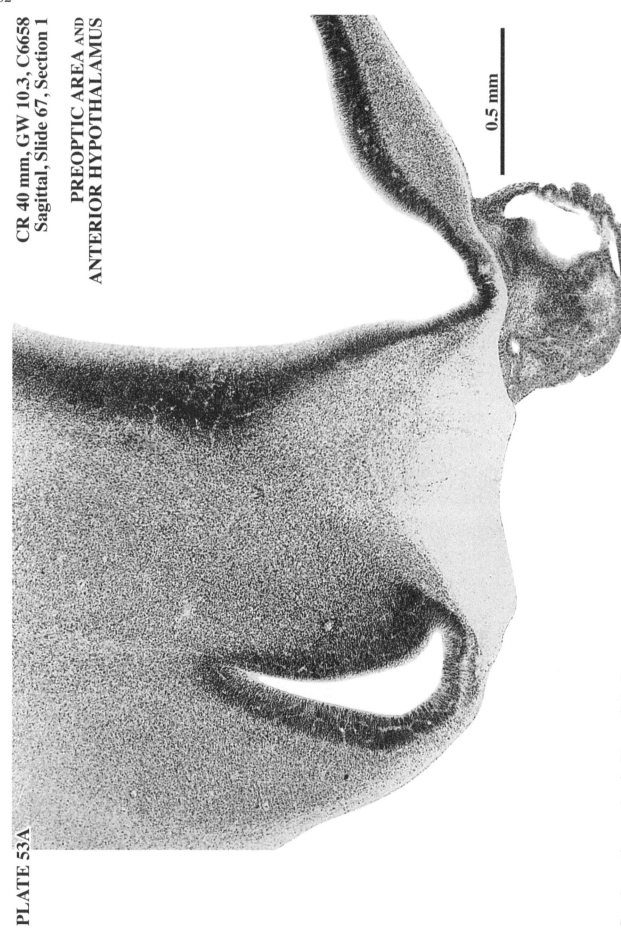

PLATE 53A

CR 40 mm, GW 10.3, C6658
Sagittal, Slide 67, Section 1

PREOPTIC AREA AND
ANTERIOR HYPOTHALAMUS

0.5 mm

See the entire section in Plates 41A–D.

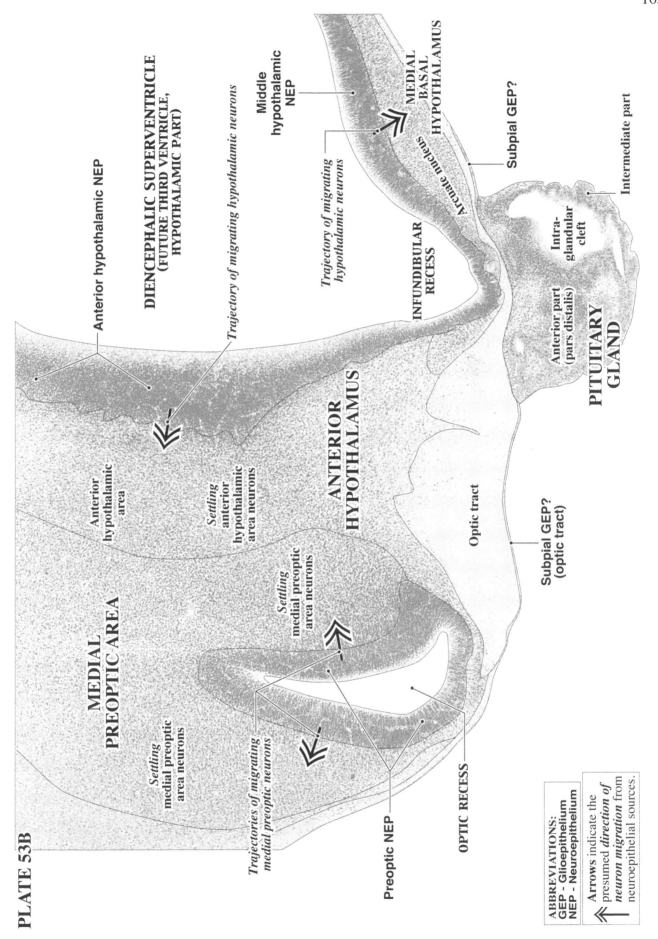

PLATE 53B

Anterior hypothalamic NEP

DIENCEPHALIC SUPERVENTRICLE
(FUTURE THIRD VENTRICLE, HYPOTHALAMIC PART)

Trajectory of migrating hypothalamic neurons

Middle hypothalamic NEP

Trajectory of migrating hypothalamic neurons

MEDIAL BASAL HYPOTHALAMUS

Arcuate nucleus

INFUNDIBULAR RECESS

Subpial GEP?

Intra-glandular cleft

Intermediate part

MEDIAL PREOPTIC AREA

Anterior hypothalamic area

Settling anterior hypothalamic area neurons

ANTERIOR HYPOTHALAMUS

Settling medial preoptic area neurons

Settling medial preoptic area neurons

Trajectories of migrating medial preoptic neurons

Preoptic NEP

OPTIC RECESS

Optic tract

Subpial GEP? (optic tract)

Anterior part (pars distalis)

PITUITARY GLAND

ABBREVIATIONS:
GEP - Glioepithelium
NEP - Neuroepithelium

Arrows indicate the presumed *direction of neuron migration* from neuroepithelial sources.

PLATE 54A

CR 40 mm, GW 10.3, C6658
Sagittal, Slide 95, Section 1

HIPPOCAMPUS AND THALAMUS

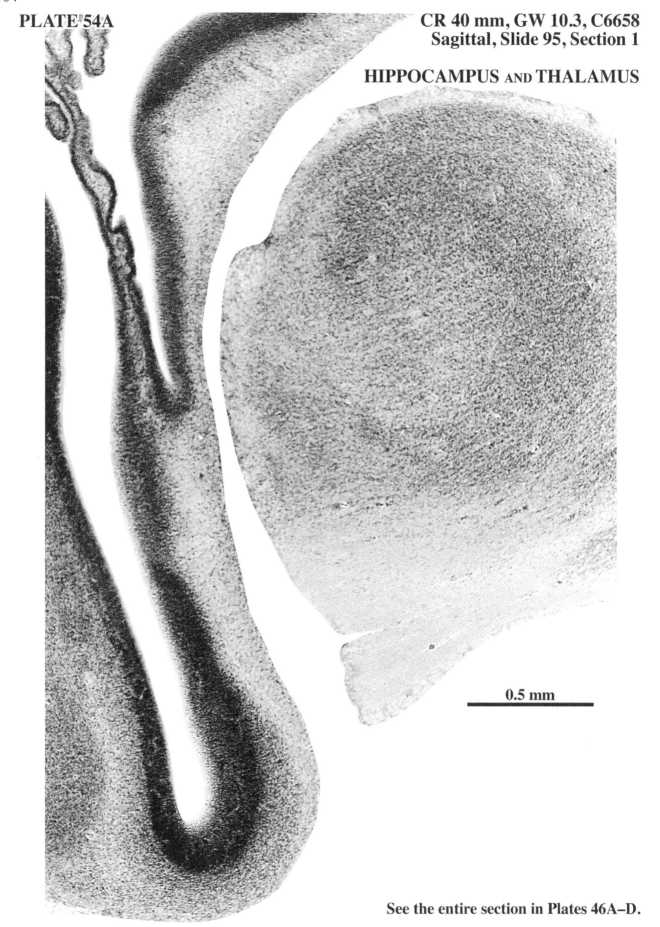

0.5 mm

See the entire section in Plates 46A–D.

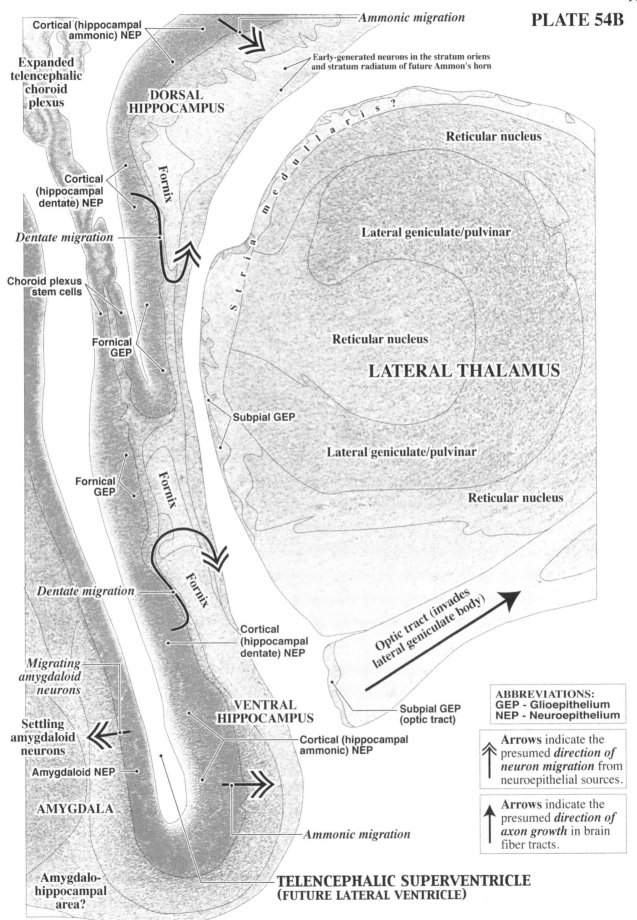

Cortical (hippocampal ammonic) NEP

Ammonic migration

Early-generated neurons in the stratum oriens and stratum radiatum of future Ammon's horn

Expanded telencephalic choroid plexus

DORSAL HIPPOCAMPUS

Reticular nucleus

Cortical (hippocampal dentate) NEP

Fornix

Lateral geniculate/pulvinar

Dentate migration

Choroid plexus stem cells

Reticular nucleus

Fornical GEP

LATERAL THALAMUS

Subpial GEP

Lateral geniculate/pulvinar

Fornical GEP

Fornix

Reticular nucleus

Dentate migration

Fornix

Cortical (hippocampal dentate) NEP

Optic tract (invades lateral geniculate body)

Migrating amygdaloid neurons

Settling amygdaloid neurons

VENTRAL HIPPOCAMPUS

Subpial GEP (optic tract)

ABBREVIATIONS:
GEP - Glioepithelium
NEP - Neuroepithelium

Cortical (hippocampal ammonic) NEP

Amygdaloid NEP

Arrows indicate the presumed *direction of neuron migration* from neuroepithelial sources.

AMYGDALA

Arrows indicate the presumed *direction of axon growth* in brain fiber tracts.

Ammonic migration

Amygdalo-hippocampal area?

TELENCEPHALIC SUPERVENTRICLE
(FUTURE LATERAL VENTRICLE)

166

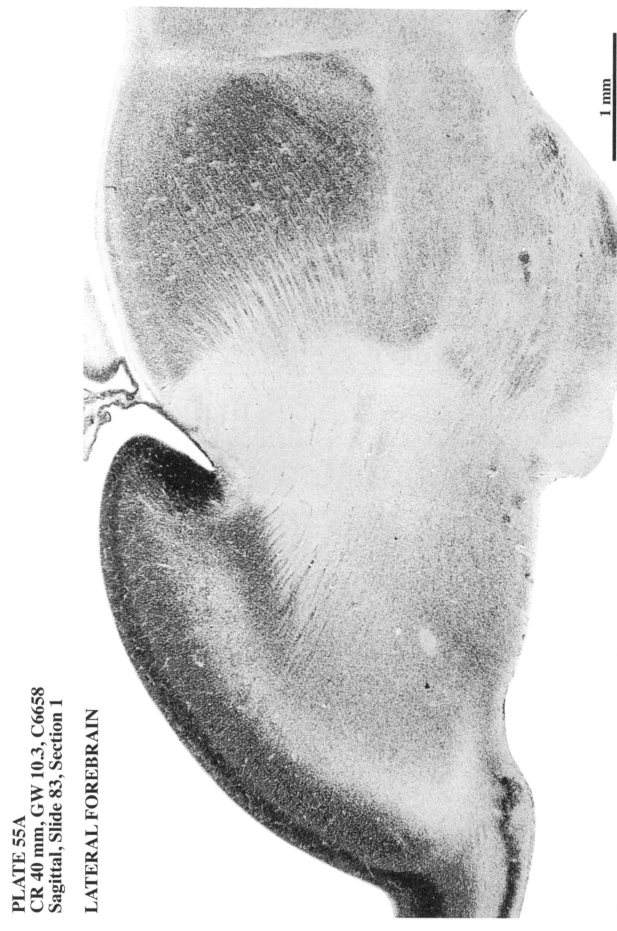

PLATE 55A
CR 40 mm, GW 10.3, C6658
Sagittal, Slide 83, Section 1

LATERAL FOREBRAIN

1 mm

See the entire section in Plates 44A–D.

PLATE 55B

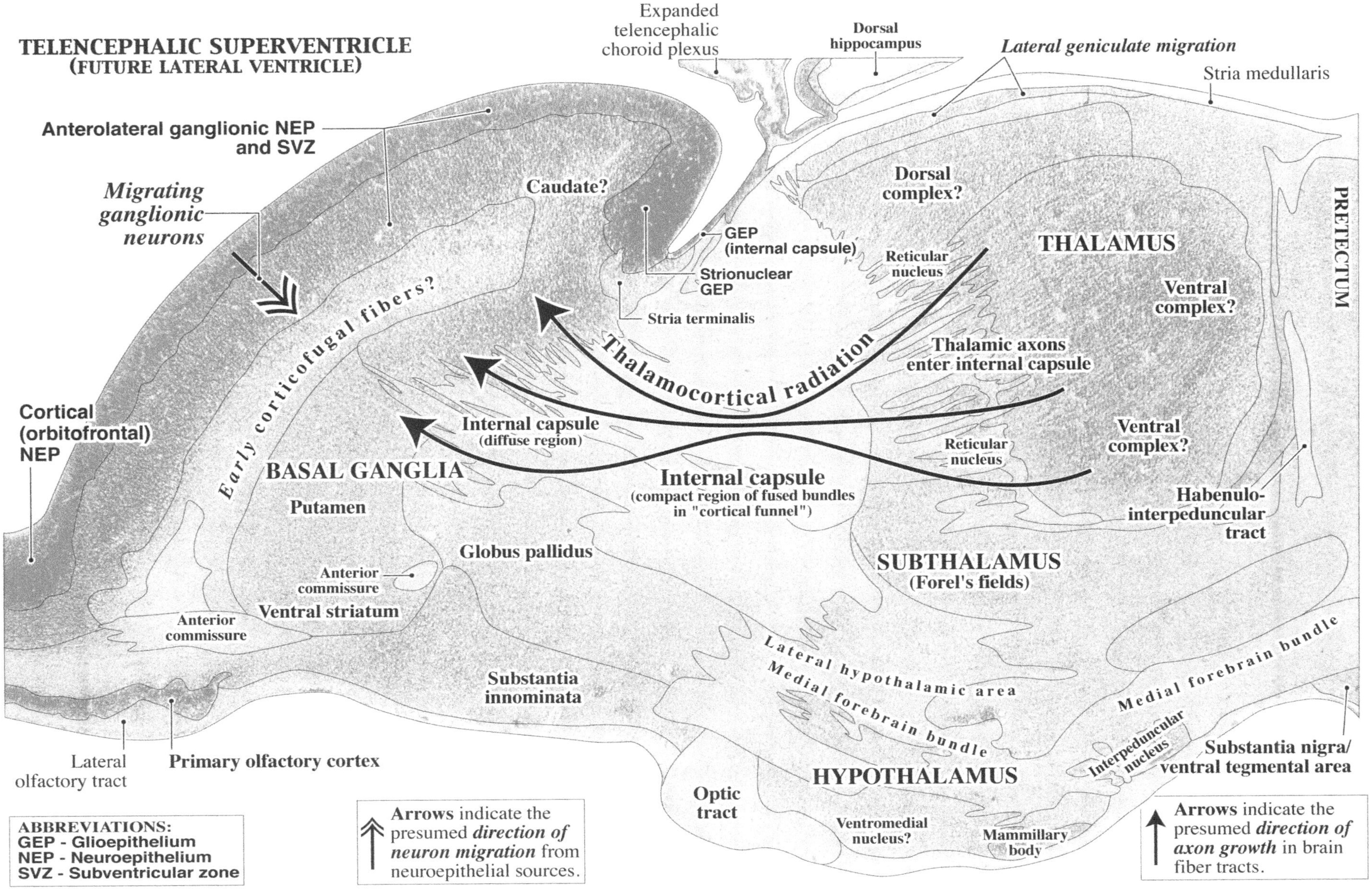

TELENCEPHALIC SUPERVENTRICLE
(FUTURE LATERAL VENTRICLE)

PRETECTUM

Stria medullaris

Lateral geniculate migration

THALAMUS

Ventral complex?

Dorsal complex?

Dorsal hippocampus

Ventral complex?

Habenulo-interpeduncular tract

Medial forebrain bundle

Substantia nigra/ventral tegmental area

Interpeduncular nucleus

Mammillary body

Reticular nucleus

Thalamic axons enter internal capsule

Reticular nucleus

Expanded telencephalic choroid plexus

GEP (internal capsule)

Strionuclear GEP

Stria terminalis

SUBTHALAMUS
(Forel's fields)

HYPOTHALAMUS

Ventromedial nucleus?

Lateral hypothalamic area

Medial forebrain bundle

Thalamocortical radiation

Caudate?

Anterolateral ganglionic NEP and SVZ

Internal capsule
(diffuse region)

Internal capsule
(compact region of fused bundles in "cortical funnel")

Globus pallidus

Optic tract

Migrating ganglionic neurons

Early corticofugal fibers?

BASAL GANGLIA

Putamen

Anterior commissure

Ventral striatum

Substantia innominata

Cortical (orbitofrontal) NEP

Anterior commissure

Primary olfactory cortex

Lateral olfactory tract

Arrows indicate the presumed *direction of axon growth* in brain fiber tracts.

Arrows indicate the presumed *direction of neuron migration* from neuroepithelial sources.

ABBREVIATIONS:
GEP - Glioepithelium
NEP - Neuroepithelium
SVZ - Subventricular zone

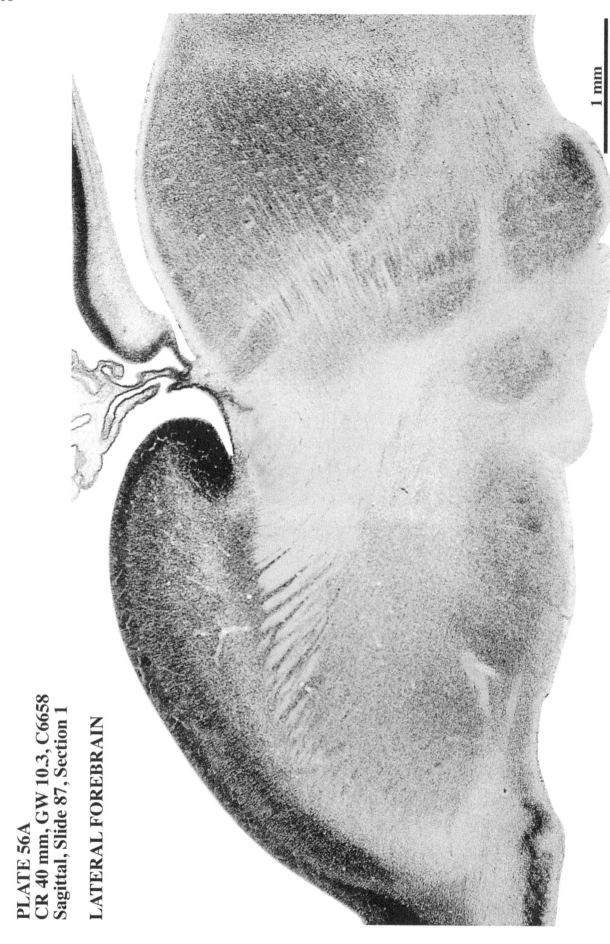

1 mm

See a low-magnification view of nearby sections in Plates 45A–D and 46A–D.

169

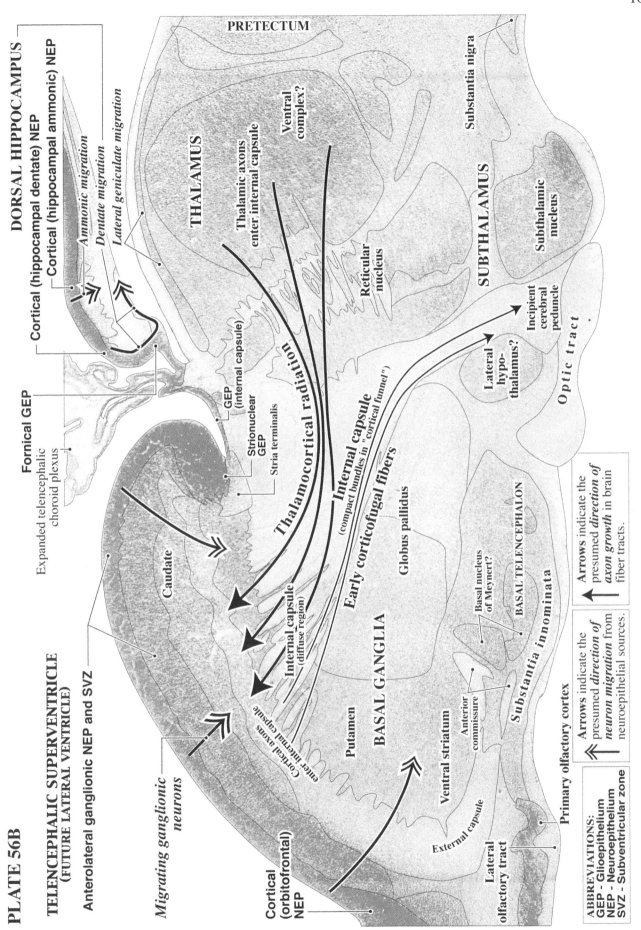

PLATE 56B

DORSAL HIPPOCAMPUS

Cortical (hippocampal dentate) NEP
Cortical (hippocampal ammonic) NEP

Ammonic migration
Dentate migration
Lateral geniculate migration

Fornical GEP

Expanded telencephalic
choroid plexus

TELENCEPHALIC SUPERVENTRICLE
(FUTURE LATERAL VENTRICLE)

Anterolateral ganglionic NEP and SVZ

Migrating ganglionic
neurons

Cortical
(orbitofrontal)
NEP

PRETECTUM

Substantia nigra

THALAMUS

Ventral
complex?

Thalamic axons
enter internal capsule

SUBTHALAMUS

Subthalamic
nucleus

Reticular
nucleus

GEP
(internal capsule)

Strionuclear
GEP

Stria terminalis

Thalamocortical radiation

Internal capsule
(compact bundles in "cortical funnel")

Early corticofugal fibers

Internal capsule
(diffuse region)

Cortical axons
enter internal capsule

Caudate

Lateral
hypo-
thalamus?

Incipient
cerebral
peduncle

Optic tract

Globus pallidus

Basal nucleus
of Meynert?

BASAL TELENCEPHALON

BASAL GANGLIA

Putamen

Anterior
commissure

Substantia innominata

Ventral striatum

External capsule

Lateral
olfactory tract

Primary olfactory cortex

Arrows indicate the
presumed direction of axon growth in brain
fiber tracts.

Arrows indicate the
presumed direction of
neuron migration from
neuroepithelial sources.

ABBREVIATIONS:
GEP - Glioepithelium
NEP - Neuroepithelium
SVZ - Subventricular zone

PLATE 57A

CR 40 mm, GW 10.3, C6658
Sagittal, Slide 95, Section 1

LATERAL FOREBRAIN

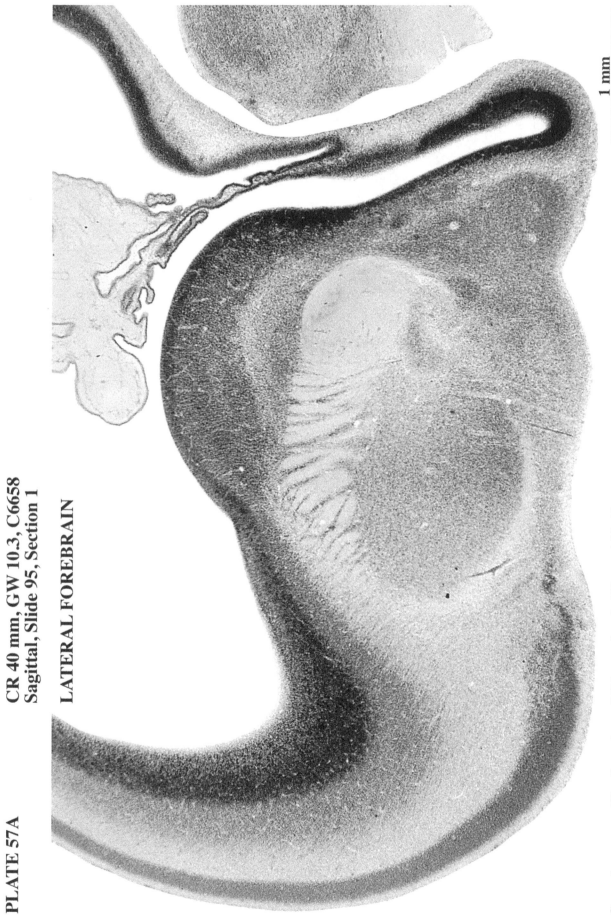

1 mm

See a low-magnification view of this section in Plates 46A–D.

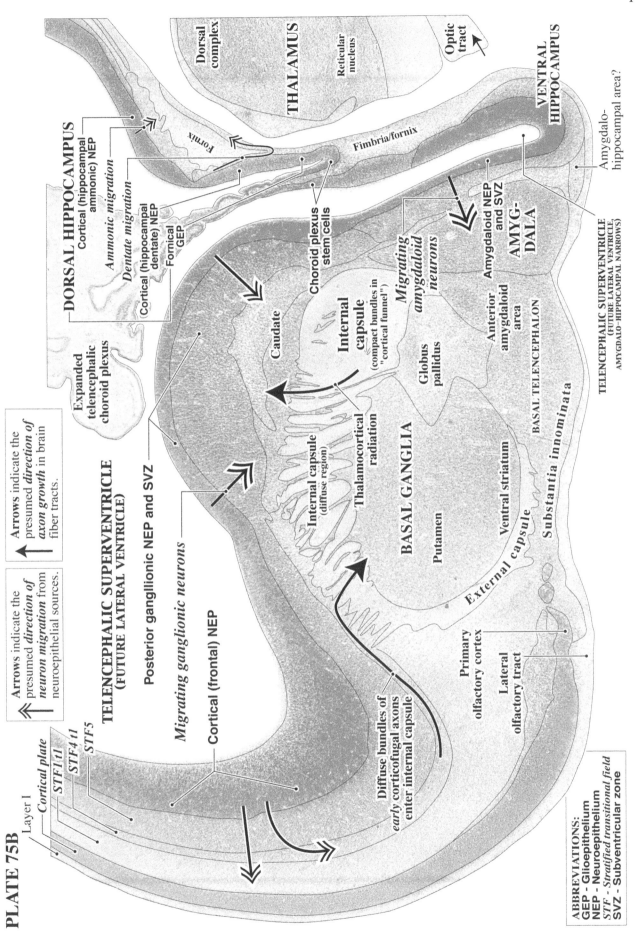

PLATE 75B

Layer 1
Cortical plate
STF1t1
STF4 t1
STF5

Arrows indicate the presumed *direction of neuron migration* from neuroepithelial sources.

Arrows indicate the presumed *direction of axon growth* in brain fiber tracts.

TELENCEPHALIC SUPERVENTRICLE
(FUTURE LATERAL VENTRICLE)

Posterior ganglionic NEP and SVZ

Migrating ganglionic neurons

Cortical (frontal) NEP

Diffuse bundles of early corticofugal axons enter internal capsule

DORSAL HIPPOCAMPUS

Cortical (hippocampal ammonic) NEP

Ammonic migration

Dentate migration

Cortical (hippocampal dentate) NEP

Fornical GEP

Expanded telencephalic choroid plexus

Caudate

Choroid plexus stem cells

Internal capsule
(compact bundles in "cortical funnel")

Migrating amygdaloid neurons

Internal capsule
(diffuse region)

Thalamocortical radiation

BASAL GANGLIA

Putamen

Globus pallidus

Ventral striatum

External capsule

Substantia innominata

Anterior amygdaloid area

AMYG-DALA

Amygdaloid NEP and SVZ

BASAL TELENCEPHALON

Primary olfactory cortex

Lateral olfactory tract

Fornix

Fimbria/fornix

THALAMUS

Dorsal complex

Reticular nucleus

Optic tract

VENTRAL HIPPOCAMPUS

Amygdalo-hippocampal area?

TELENCEPHALIC SUPERVENTRICLE
(FUTURE LATERAL VENTRICLE,
AMYGDALO–HIPPOCAMPAL NARROWS)

ABBREVIATIONS:
GEP - Glioepithelium
NEP - Neuroepithelium
STF - Stratified transitional field
SVZ - Subventricular zone

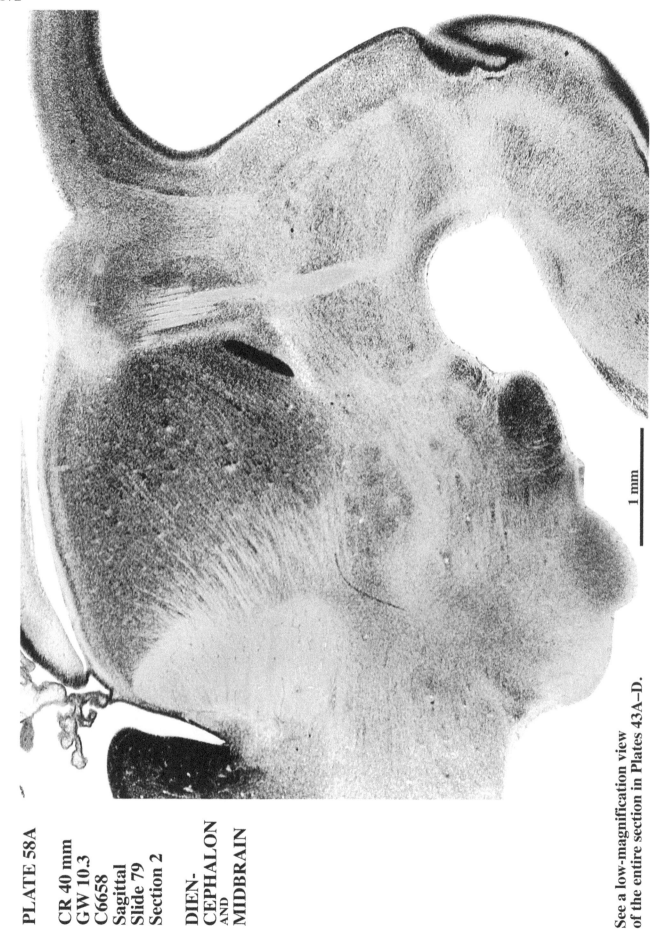

PLATE 58A

CR 40 mm
GW 10.3
C6658
Sagittal
Slide 79
Section 2

DIEN-
CEPHALON
AND
MIDBRAIN

1 mm

See a low-magnification view
of the entire section in Plates 43A–D.

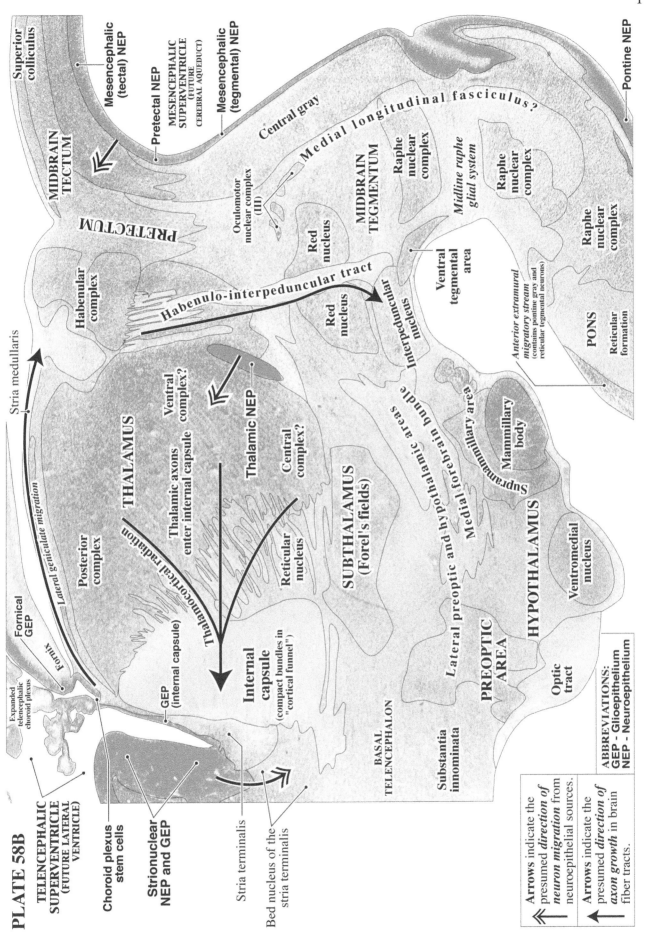

173

PLATE 58B

Superior colliculus

Mesencephalic (tectal) NEP

Pretectal NEP

MESENCEPHALIC SUPERVENTRICLE (FUTURE CEREBRAL AQUEDUCT)

Mesencephalic (tegmental) NEP

Pontine NEP

MIDBRAIN TECTUM

PRETECTUM

Central gray

Medial longitudinal fasciculus?

Oculomotor nuclear complex (III)

Red nucleus

MIDBRAIN TEGMENTUM

Raphe nuclear complex

Midline raphe glial system

Raphe nuclear complex

Habenular complex

Habenulo-interpeduncular tract

Red nucleus

Interpeduncular nucleus

Ventral tegmental area

Raphe nuclear complex

Stria medullaris

Lateral geniculate migration

THALAMUS

Ventral complex?

Thalamic axons enter internal capsule

Thalamic NEP

Central complex?

Anterior extramural migratory stream (contains pontine gray and reticular tegmental neurons)

PONS

Reticular formation

Fornical GEP

Fornix

Posterior complex

Thalamocortical radiation

Thalamic (internal capsule)

Reticular nucleus

SUBTHALAMUS (Forel's fields)

Medial forebrain bundle

Supramammillary area

Mammillary body

Expanded telencephalic choroid plexus

GEP (internal capsule)

Internal capsule (compact bundles in "cortical funnel")

Lateral preoptic and hypothalamic areas

HYPOTHALAMUS

Ventromedial nucleus

TELENCEPHALIC SUPERVENTRICLE (FUTURE LATERAL VENTRICLE)

Choroid plexus stem cells

Strionuclear NEP and GEP

Stria terminalis

Bed nucleus of the stria terminalis

BASAL TELENCEPHALON

Substantia innominata

PREOPTIC AREA

Optic tract

ABBREVIATIONS:
GEP - Glioepithelium
NEP - Neuroepithelium

Arrows indicate the presumed *direction of neuron migration* from neuroepithelial sources.

Arrows indicate the presumed *direction of axon growth* in brain fiber tracts.

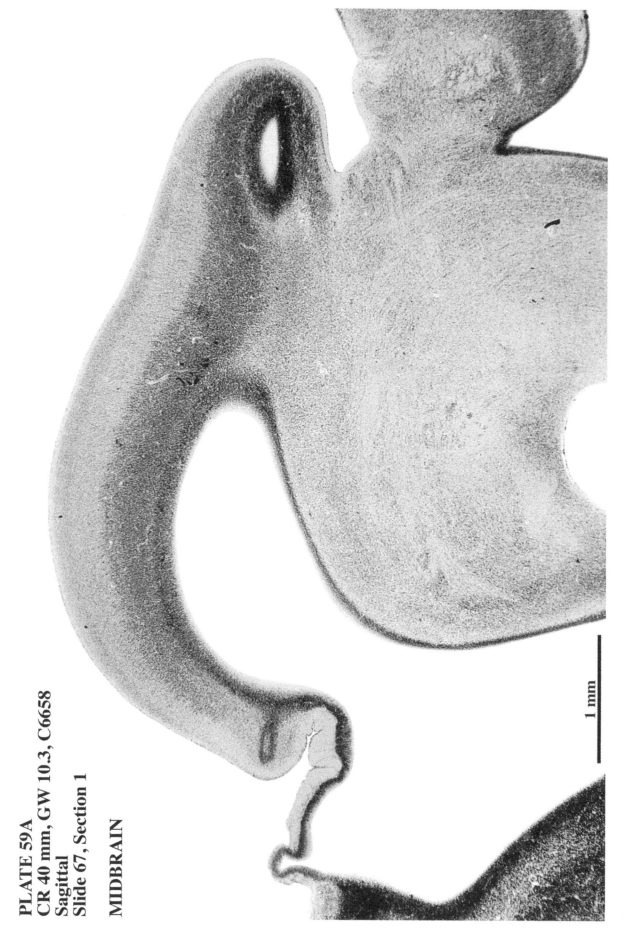

See a low-magnification view of this section in Plates 41A–D.

1 mm

PLATE 59A
CR 40 mm, GW 10.3, C6658
Sagittal
Slide 67, Section 1

MIDBRAIN

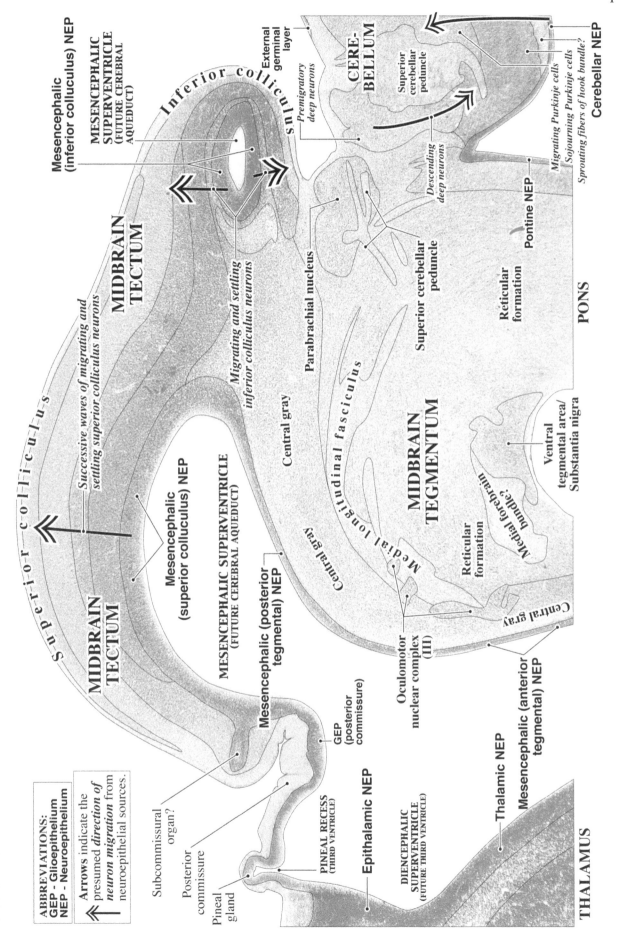

PLATE 59B

ABBREVIATIONS:
GEP - Glioepithelium
NEP - Neuroepithelium

Arrows indicate the presumed *direction of neuron migration* from neuroepithelial sources.

Mesencephalic (inferior colliculus) NEP

MESENCEPHALIC SUPERVENTRICLE (FUTURE CEREBRAL AQUEDUCT)

External germinal layer

Premigratory deep neurons

CERE-BELLUM

Inferior colliculus

Superior cerebellar peduncle

Migrating Purkinje cells
Sojourning Purkinje cells
Sprouting fibers of hook bundle?

Cerebellar NEP

MIDBRAIN TECTUM

MIDBRAIN TECTUM

Descending deep neurons

Successive waves of migrating and settling superior colliculus neurons

Superior colliculus

Migrating and settling inferior colliculus neurons

Parabrachial nucleus

Superior cerebellar peduncle

Pontine NEP

Reticular formation

PONS

Mesencephalic (superior colliculus) NEP

Central gray

Central gray

MESENCEPHALIC SUPERVENTRICLE (FUTURE CEREBRAL AQUEDUCT)

Medial longitudinal fasciculus

Medial longitudinal fasciculus

MIDBRAIN TEGMENTUM

Reticular formation

Medial forebrain bundle?

Ventral tegmental area/ Substantia nigra

Mesencephalic (posterior tegmental) NEP

GEP (posterior commissure)

Oculomotor nuclear complex (III)

Central gray

Epithalamic NEP

Subcommissural organ?

Posterior commissure

Pineal gland

PINEAL RECESS (THIRD VENTRICLE)

DIENCEPHALIC SUPERVENTRICLE (FUTURE THIRD VENTRICLE)

Thalamic NEP

Mesencephalic (anterior tegmental) NEP

THALAMUS

175

176

CR 40 mm, GW 10.3, C6658
Sagittal, Slide 63, Section 1

MIDBRAIN TECTUM

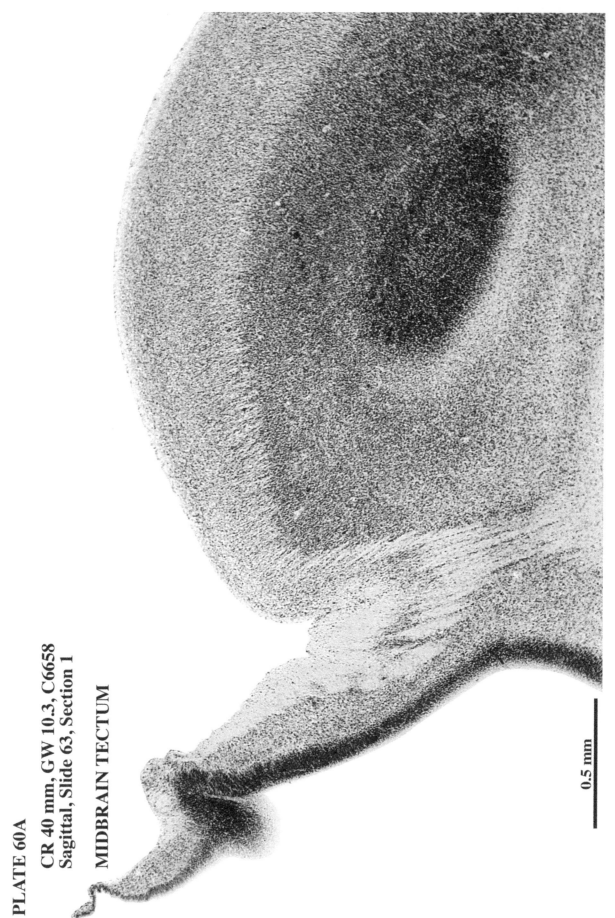

See a low-magnification view of this section in Plates 40A–D.

0.5 mm

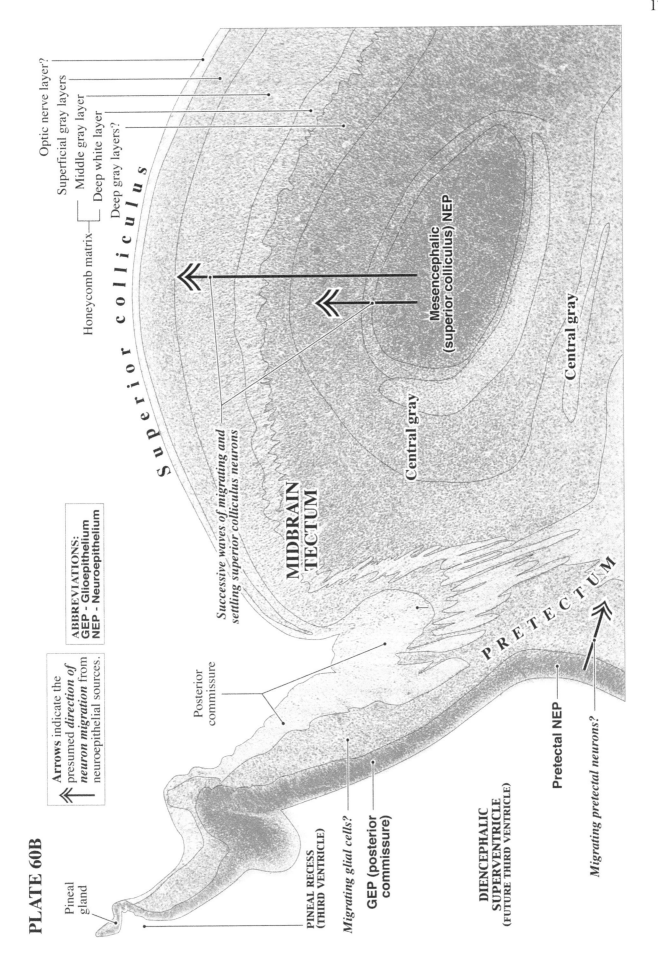

PLATE 60B

177

Pineal gland

Optic nerve layer?
Superficial gray layers
Middle gray layer
Deep white layer
Deep gray layers?

Honeycomb matrix

s u p e r i o r c o l l i c u l u s

Successive waves of migrating and settling superior colliculus neurons

MIDBRAIN TECTUM

Central gray

Central gray

Mesencephalic (superior colliculus) NEP

ABBREVIATIONS:
GEP - Glioepithelium
NEP - Neuroepithelium

Arrows indicate the presumed *direction of neuron migration* from neuroepithelial sources.

Posterior commissure

PINEAL RECESS (THIRD VENTRICLE)

Migrating glial cells?

GEP (posterior commissure)

DIENCEPHALIC SUPERVENTRICLE (FUTURE THIRD VENTRICLE)

P R E T E C T U M

Pretectal NEP

Migrating pretectal neurons?

PLATE 61A

CR 40 mm, GW 10.3, C6658
Sagittal, Slide 71, Section 2
CEREBELLUM

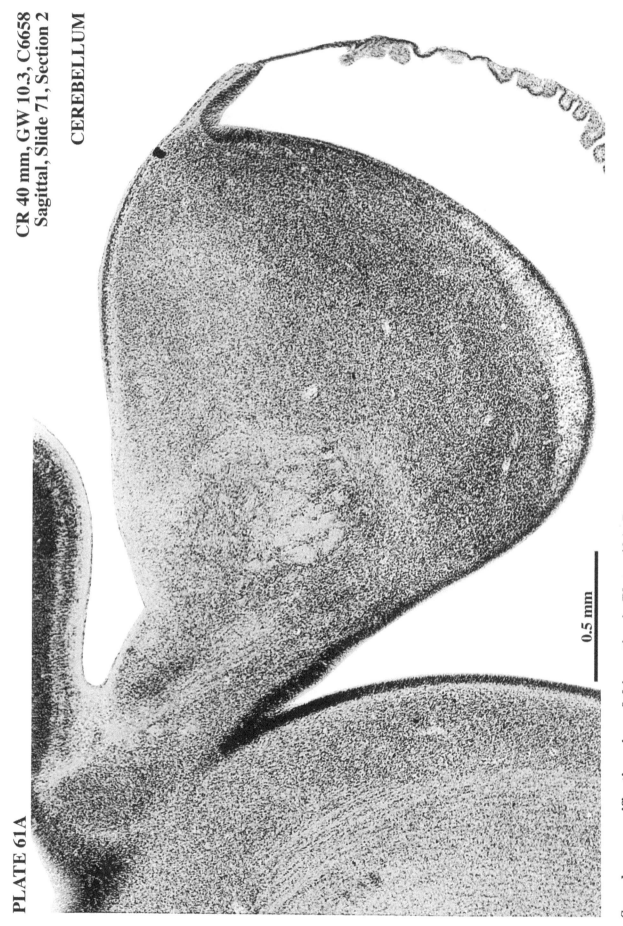

0.5 mm

See a low-magnification view of this section in Plates 42A–D.

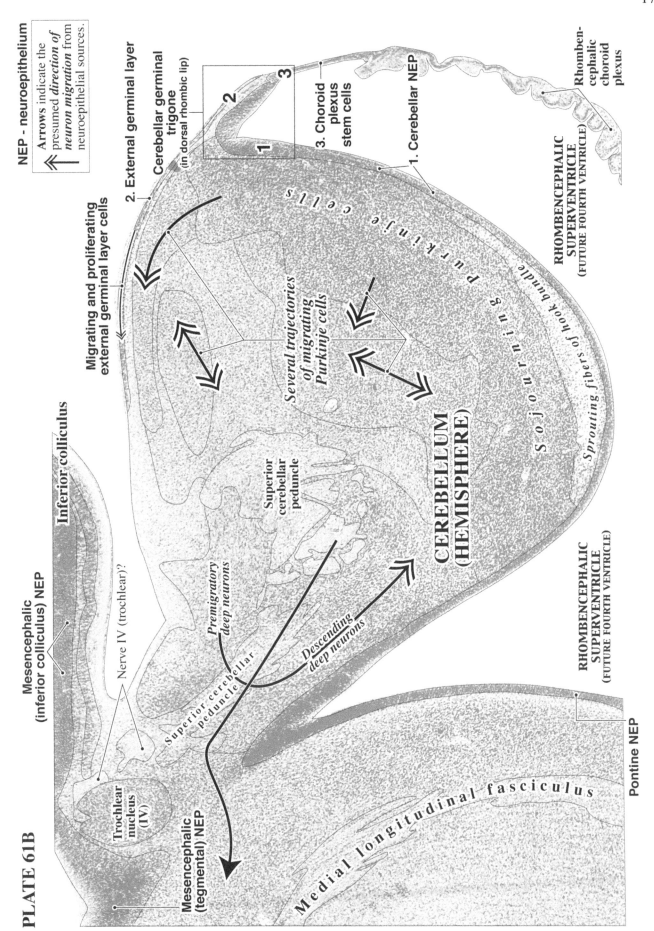

PLATE 61B

NEP - neuroepithelium

Arrows indicate the presumed *direction of neuron migration* from neuroepithelial sources.

2. External germinal layer

Cerebellar germinal trigone
(in dorsal rhombic lip)

3. Choroid plexus stem cells

1. Cerebellar NEP

Cerebellar NEP

Rhomben-cephalic choroid plexus

RHOMBENCEPHALIC SUPERVENTRICLE
(FUTURE FOURTH VENTRICLE)

Sojourning Purkinje cells

Migrating and proliferating external germinal layer cells

Several trajectories of migrating Purkinje cells

Sprouting fibers of hook bundle

Inferior colliculus

CEREBELLUM (HEMISPHERE)

Mesencephalic (inferior colliculus) NEP

Superior cerebellar peduncle

Nerve IV (trochlear)?

Premigratory deep neurons

Superior cerebellar peduncle

Descending deep neurons

RHOMBENCEPHALIC SUPERVENTRICLE
(FUTURE FOURTH VENTRICLE)

Trochlear nucleus (IV)

Mesencephalic (tegmental) NEP

Medial longitudinal fasciculus

Pontine NEP

PLATE 62A
CR 40 mm, GW 10.3, C6658
Sagittal, Slide 75, Section 2

PONS/MEDULLA

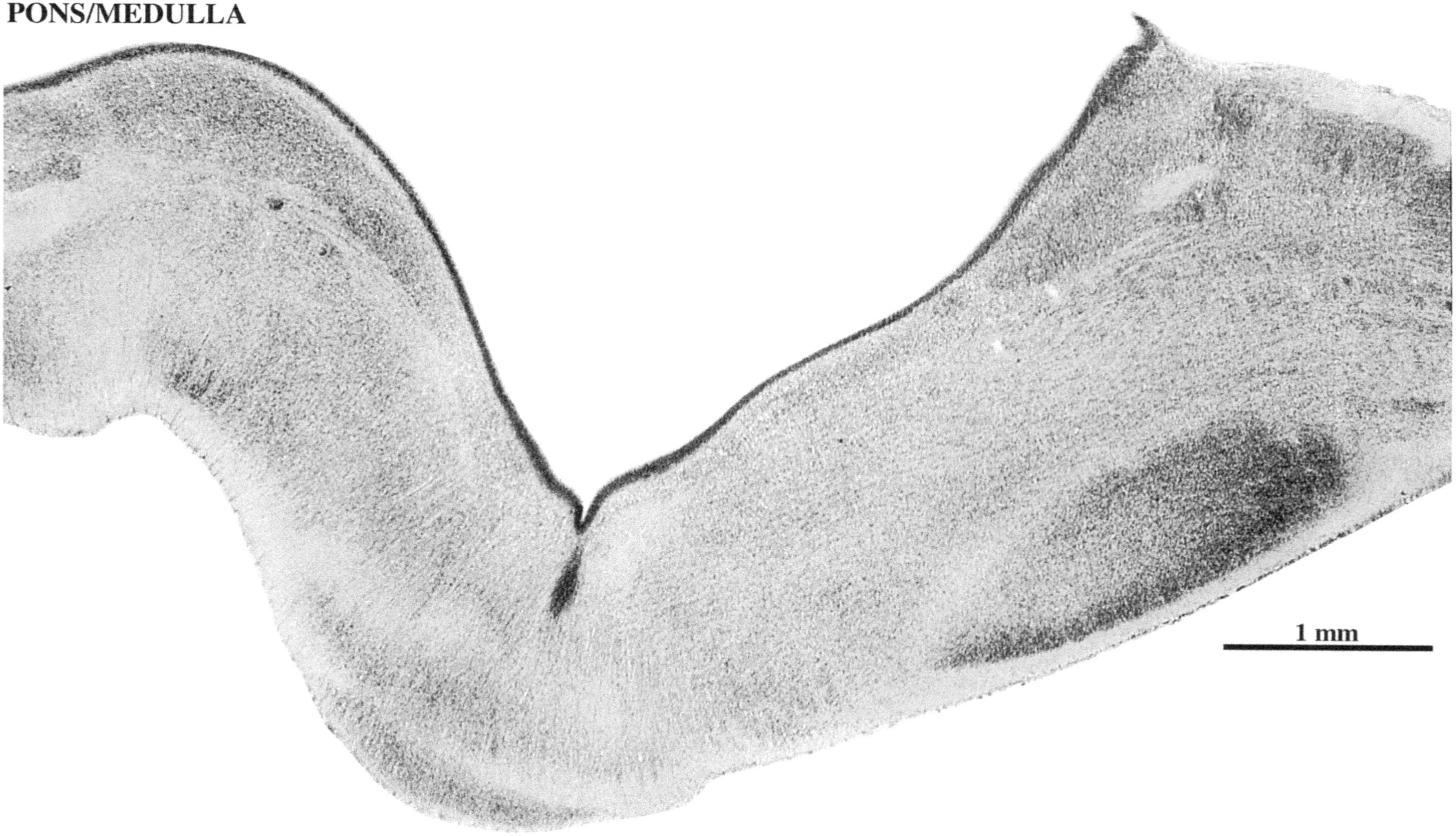

1 mm

See a low-magnification view of this section in Plates 43A–D.

PLATE 62B

NEP - neuroepithelium

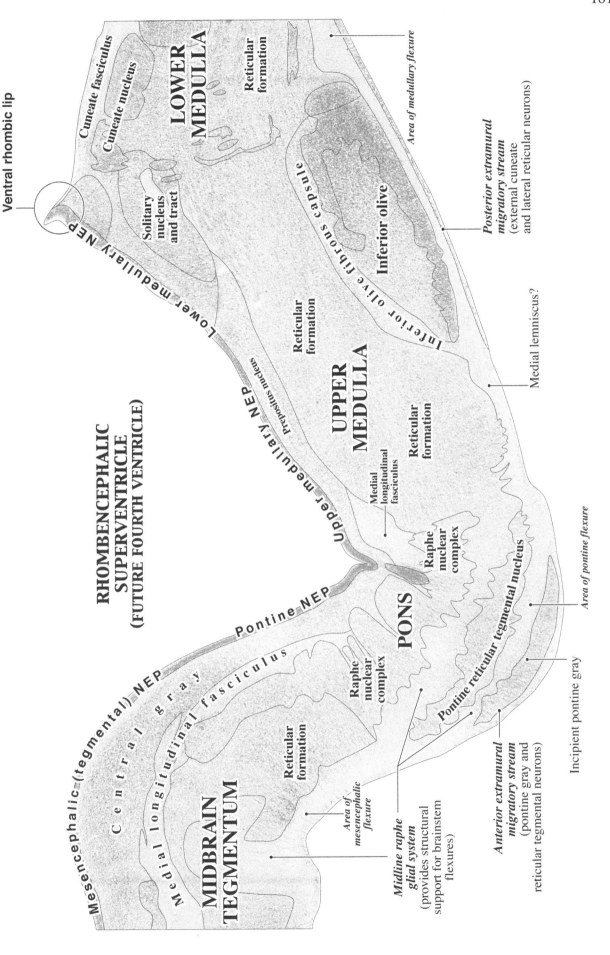

Ventral rhombic lip

LOWER MEDULLA

Cuneate fasciculus

Cuneate nucleus

Reticular formation

Area of medullary flexure

Solitary nucleus and tract

Lower medullary NEP

Posterior extramural migratory stream (external cuneate and lateral reticular neurons)

Inferior olive

Inferior olive fibrous capsule

Medial lemniscus?

Reticular formation

UPPER MEDULLA

Reticular formation

Upper medullary NEP

Prepositus nucleus

RHOMBENCEPHALIC SUPERVENTRICLE (FUTURE FOURTH VENTRICLE)

Medial longitudinal fasciculus

Raphe nuclear complex

Pontine NEP

PONS

Raphe nuclear complex

Area of pontine flexure

Pontine reticular tegmental nucleus

Incipient pontine gray

Mesencephalic (tegmental) NEP

Central gray

Medial longitudinal fasciculus

Reticular formation

MIDBRAIN TEGMENTUM

Area of mesencephalic flexure

Midline raphe glial system (provides structural support for brainstem flexures)

Anterior extramural migratory stream (pontine gray and reticular tegmental neurons)

PLATE 63A

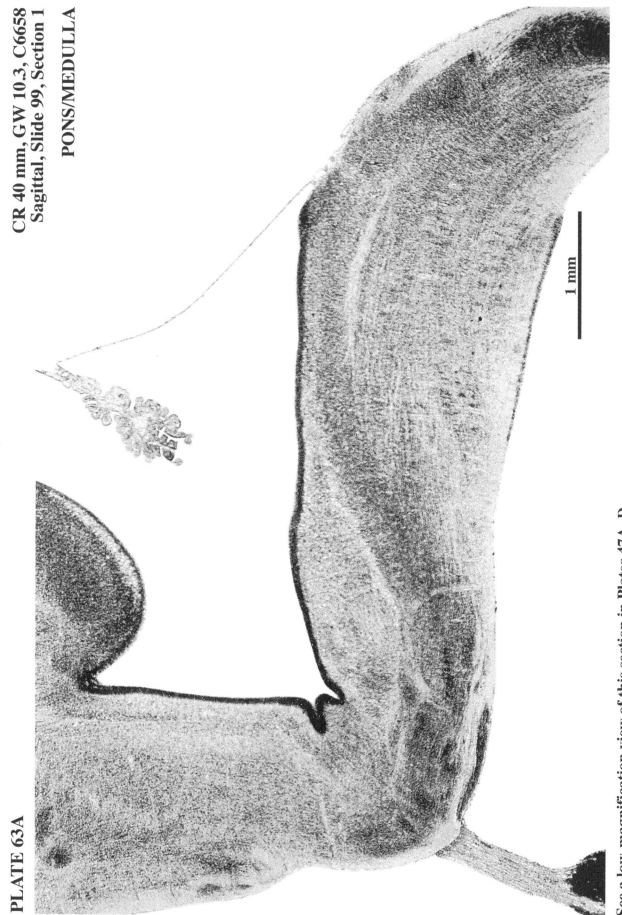

CR 40 mm, GW 10.3, C6658
Sagittal, Slide 99, Section 1
PONS/MEDULLA

1 mm

See a low-magnification view of this section in Plates 47A–D.

NEP - neuroepithelium

PLATE 63B

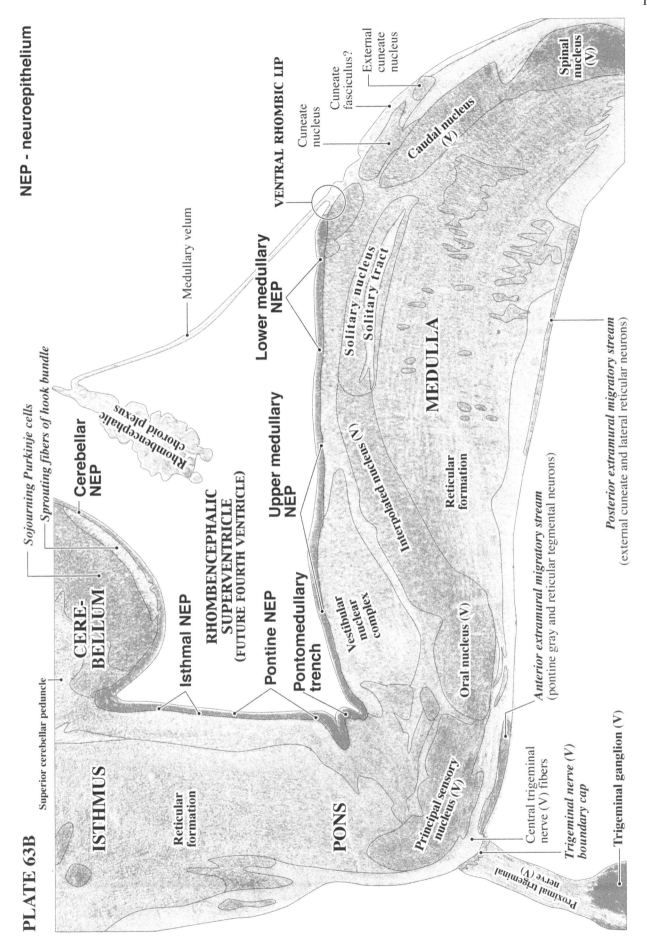

Superior cerebellar peduncle

Sojourning Purkinje cells
Sprouting fibers of hook bundle

Cerebellar NEP

Medullary velum

Rhombencephalic choroid plexus

CERE-BELLUM

Isthmal NEP

RHOMBENCEPHALIC SUPERVENTRICLE (FUTURE FOURTH VENTRICLE)

Pontine NEP

Upper medullary NEP

Lower medullary NEP

VENTRAL RHOMBIC LIP

Cuneate fasciculus?
External cuneate nucleus

Cuneate nucleus

Spinal nucleus (V)

Caudal nucleus (V)

Pontomedullary trench

ISTHMUS

Reticular formation

Vestibular nuclear complex

Solitary nucleus
Solitary tract

Interpolated nucleus (V)

MEDULLA

Reticular formation

Oral nucleus (V)

Anterior extramural migratory stream (pontine gray and reticular tegmental neurons)

Posterior extramural migratory stream (external cuneate and lateral reticular neurons)

PONS

Principal sensory nucleus (V)

Central trigeminal nerve (V) fibers

Trigeminal nerve (V) boundary cap

Proximal trigeminal nerve (V)

Trigeminal ganglion (V)

PLATE 64A

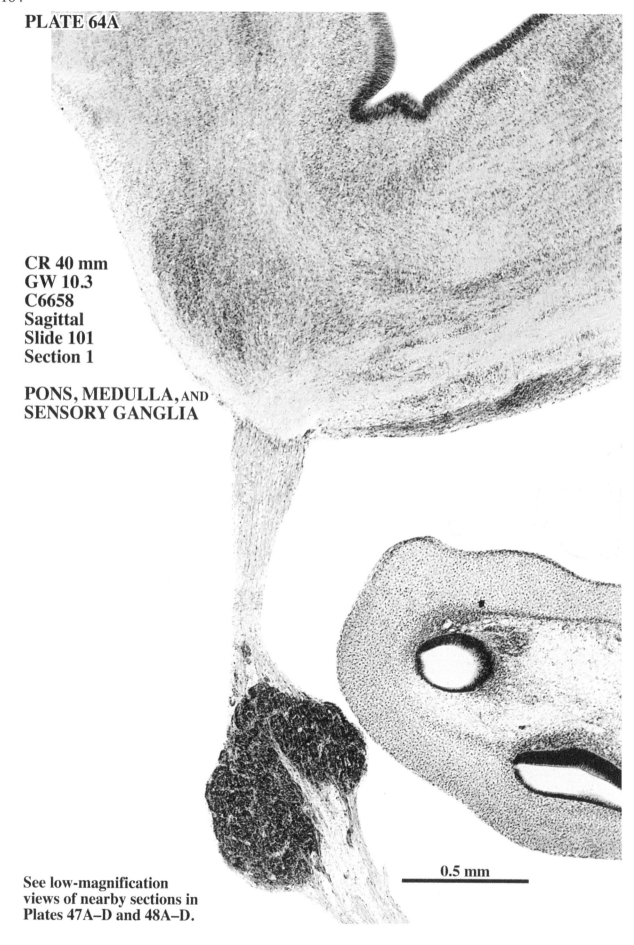

CR 40 mm
GW 10.3
C6658
Sagittal
Slide 101
Section 1

PONS, MEDULLA, AND
SENSORY GANGLIA

0.5 mm

See low-magnification
views of nearby sections in
Plates 47A–D and 48A–D.

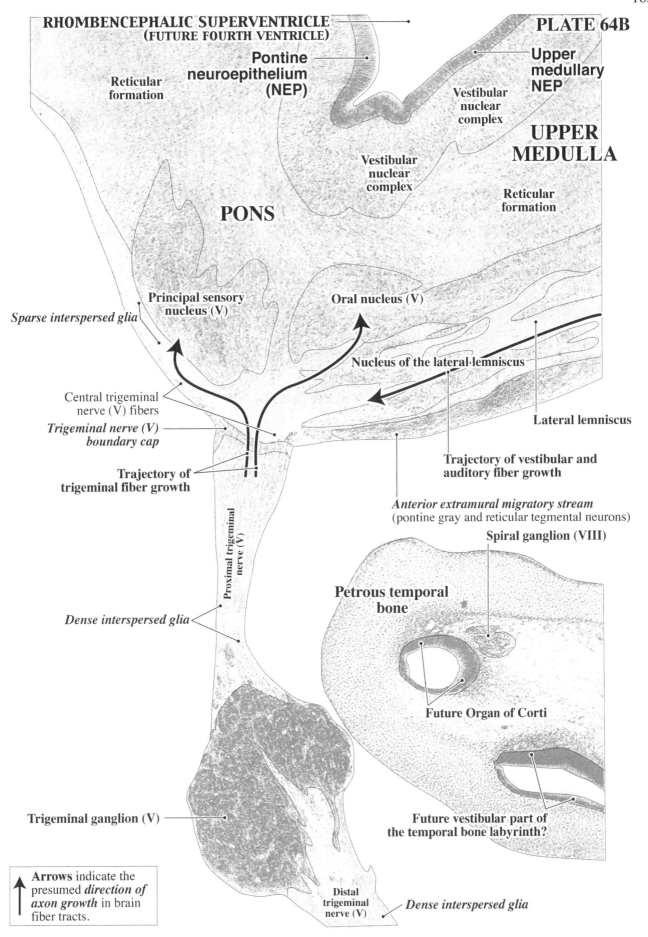

RHOMBENCEPHALIC SUPERVENTRICLE
(FUTURE FOURTH VENTRICLE)

Reticular
formation

Pontine
neuroepithelium
(NEP)

Upper
medullary
NEP

Vestibular
nuclear
complex

**UPPER
MEDULLA**

Vestibular
nuclear
complex

Reticular
formation

PONS

Principal sensory
nucleus (V)

Oral nucleus (V)

Sparse interspersed glia

Nucleus of the lateral lemniscus

Central trigeminal
nerve (V) fibers

*Trigeminal nerve (V)
boundary cap*

Lateral lemniscus

Trajectory of
trigeminal fiber growth

**Trajectory of vestibular and
auditory fiber growth**

Anterior extramural migratory stream
(pontine gray and reticular tegmental neurons)

Proximal trigeminal nerve (V)

Spiral ganglion (VIII)

Dense interspersed glia

**Petrous temporal
bone**

Future Organ of Corti

Trigeminal ganglion (V)

**Future vestibular part of
the temporal bone labyrinth?**

Arrows indicate the
presumed *direction of
axon growth* in brain
fiber tracts.

Distal
trigeminal
nerve (V)

Dense interspersed glia

186

PLATE 65A

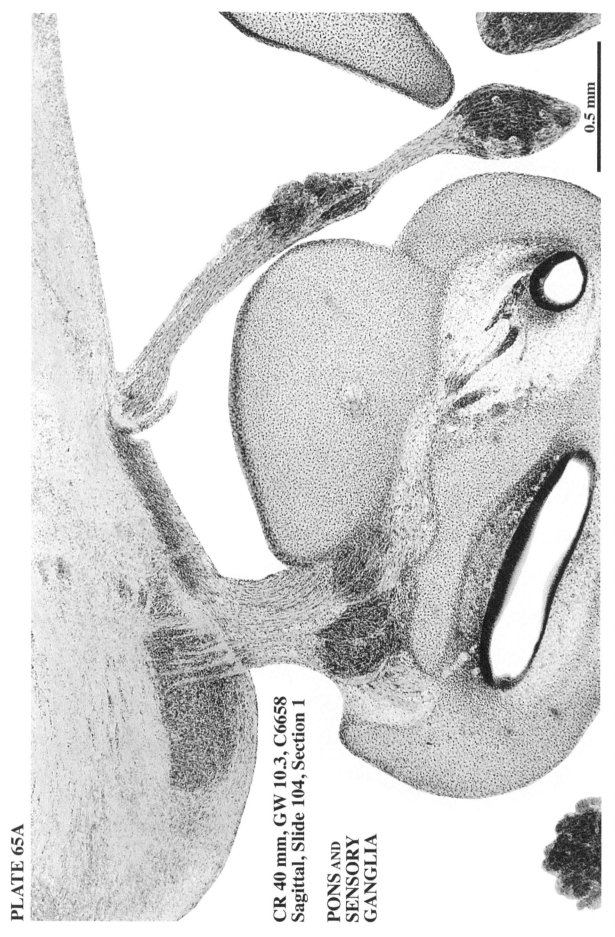

CR 40 mm, GW 10.3, C6658
Sagittal, Slide 104, Section 1

PONS AND
SENSORY
GANGLIA

0.5 mm

See low-magnification views of nearby sections in Plates 48A–D and 49A–D.

PLATE 65B

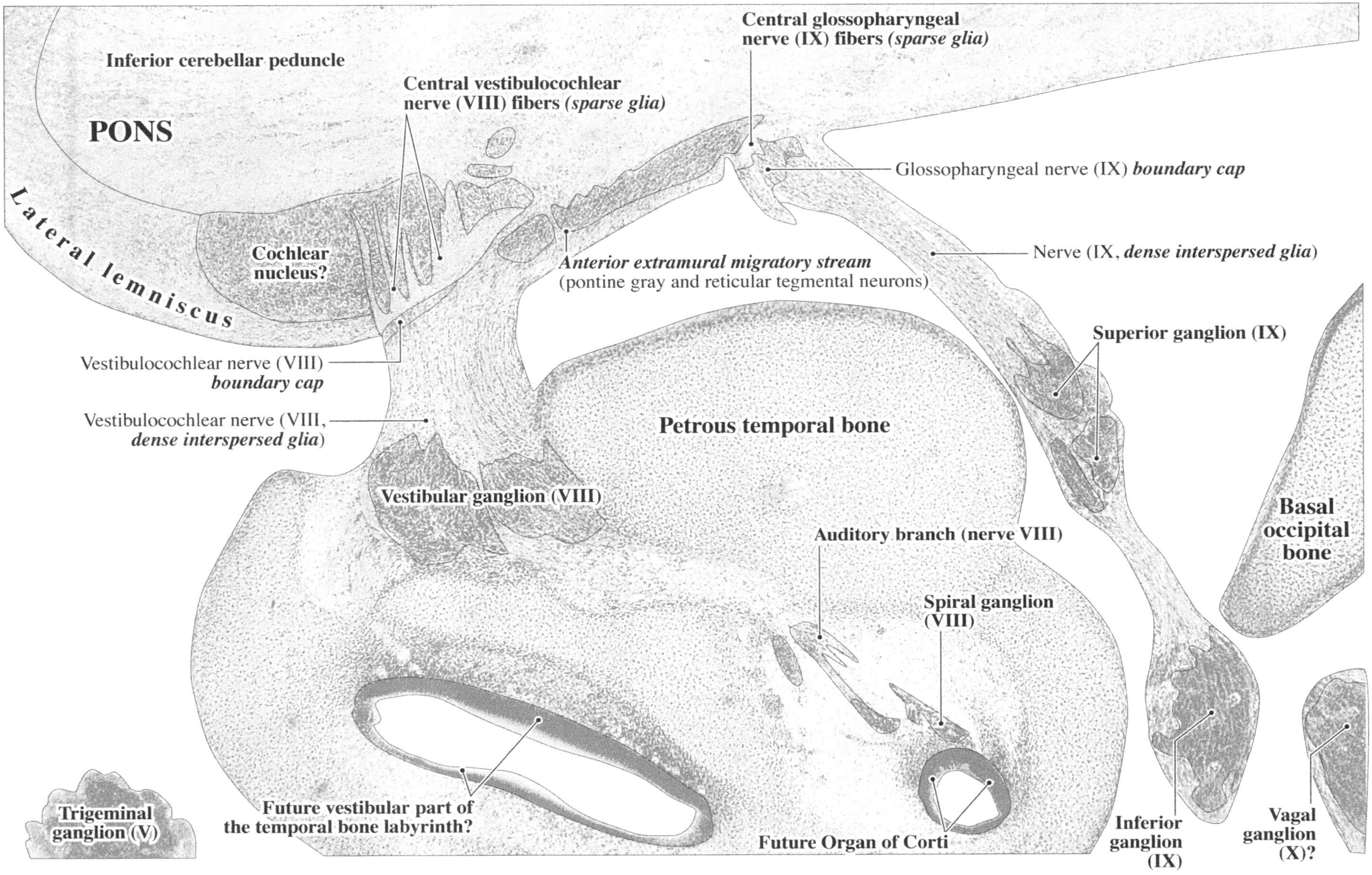

Inferior cerebellar peduncle

PONS

L a t e r a l   l e m n i s c u s

Central glossopharyngeal
nerve (IX) fibers (*sparse glia*)

Glossopharyngeal nerve (IX) *boundary cap*

Nerve (IX, *dense interspersed glia*)

Superior ganglion (IX)

Basal
occipital
bone

Vagal
ganglion
(X)?

Inferior
ganglion
(IX)

Anterior extramural migratory stream
(pontine gray and reticular tegmental neurons)

Central vestibulocochlear
nerve (VIII) fibers (*sparse glia*)

Auditory branch (nerve VIII)

Spiral ganglion
(VIII)

Petrous temporal bone

Future Organ of Corti

Cochlear
nucleus?

Vestibulocochlear nerve (VIII)
boundary cap

Vestibulocochlear nerve (VIII,
dense interspersed glia)

Vestibular ganglion (VIII)

Future vestibular part of
the temporal bone labyrinth?

Trigeminal
ganglion (V)

9 781032 219370